CARTRIDGES

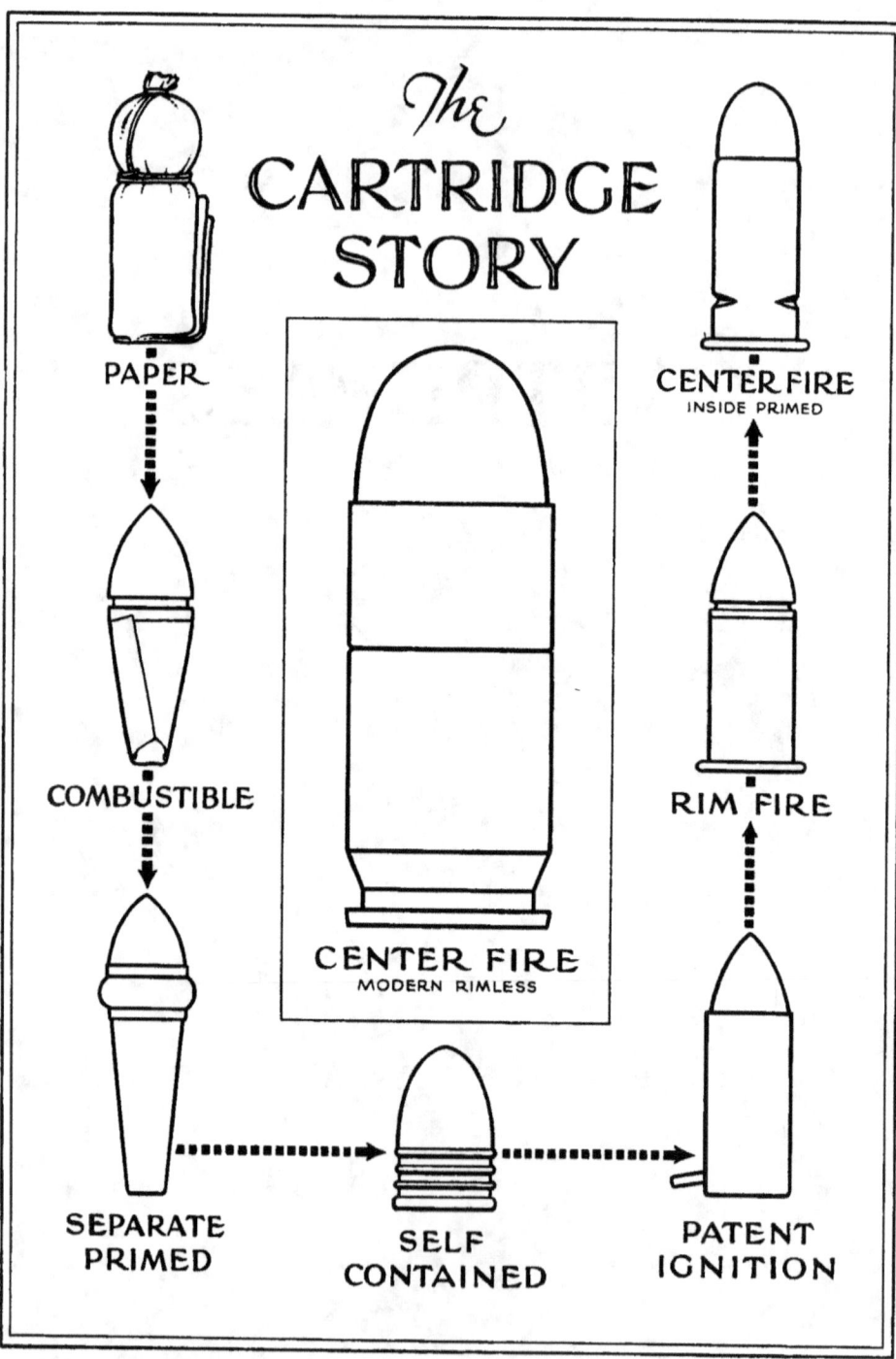

CARTRIDGES
-a pictorial digest of small arms ammunition

by

Herschel C. Logan
AUTHOR AND ILLUSTRATOR
HAND CANNON TO AUTOMATIC

BONANZA BOOKS · NEW YORK

©Copyright MCMLIX by
The Stackpole Company

This edition published by Bonanza Books,
a division of Crown Publishers, Inc.,
by arrangement with The Stackpole Company

A PATHFINDER BOOKS REPRINT EDITION
Printed in the United States of America
ISBN: 978-1951682408

To
the Memory of
OLIVER C. LOGAN
My Father

thank you

No book of this nature can be the work of one person. It is only through the kindness, cooperation and assistance of collectors, dealers and friends, that such a study can be accomplished.

Mere words are most inadequate in expressing my deep sense of appreciation and gratitude to all who have helped make possible the compiling of this digest. Many have loaned rare specimens from their personal collections from which to make the drawings. Others have supplied photographs, detailed specifications, data and suggestions.

Special recognition should go to the following whose assistance has been most valuable and helpful . . .

Raymond V. Alquist
John F. Baker
George W. Barnes
Miller Bedford
Herb Brand
Fred J. Braucher
Charles Edward Chapel
Herman P. Dean
F. Theodore Dexter
Albert Foster, Jr.
Lt. Col. Calvin Goddard
James J. Grant
Alton Jones
Elmer Keith
W. G. C. Kimball
Major B. R. Lewis
Sam Logan
Philip Jay Medicus
Philip Medicus

Carl Metzger
Stuart Miller
Platt Monfort
B. D. Munhall
A. Murphy
Ray Riling
A. W. Rowe
Shiff the Gunman
W. H. B. Smith
James E. Serven
Frank Slack
Maj. Hugh Smiley
J. M. Standish
Henry M. Stewart
Jerald T. Teesdale
Frank Wheeler
Frank Williams
Clyde Wilson
Eldon G. Wolff

So, again I say . . . THANK YOU . . . and it's from the heart.

foreword

THIS BOOK is designed to be a companion book to *Hand Cannon to Automatic*. So many favorable comments have been received upon the style of presentation in the previous work that a similar method has been employed in this digest of the interesting development of cartridges.

For the most part brief descriptions have been used along with the actual size illustrations of the cartridges. Technical terms have been avoided, since this is not a book primarily for identification. Velocity, trajectory, ballistics and other technical phases of cartridges have been left to experts better qualified for such scientific study. Detailed "miked" measurements have been reduced to a minimum. Only simple measurements to the nearest thirty-second of an inch have been used in designating case lengths. Many collectors and laymen have expressed a desire for a pictorial presentation of cartridges. It is hoped that the present volume will merit their approval. One feature which should be of especial value is the CARTRIDGE CHRONOLOGY. Taken from generally accepted dates, it gives briefly the graphic story of cartridge development.

Right here let it be said that inaccuracies and discrepancies may be found. That is recognized. But with a compilation of this nature, gleaned from many sources both official and otherwise, such things will occur. However, it is hoped that they will not be too numerous and that the book may spur others into works of a more comprehensive nature. It is regrettable that in the past there has been so little information available on cartridges. One wonders why this inseparable companion of arms has been so neglected . . . even by arms writers. What could be more useless than a military or sporting arm . . . without ammunition?

The book has been divided into sections . . . each a vital part of the story, yet each distinctive in itself. No effort has been made to subdivide hand gun and long arm cartridges. In many instances the same cartridge was used in both. To divide them, then, would only add confusion. The only deviation from a chronological order is in the method of starting with the smaller caliber first and proceeding on through to the larger caliber in each section. This order should make for easier reference.

To select the cartridges pictured herein, out of the thousands of types and variations, has not been an easy task. In addition to the historical types there have been added many other items which are of unusual interest to collectors, and which it is hoped will serve to further visualize the development story. In many instances more than one caliber of a series is illustrated. Many rare specimens are pictured herein for the first time. Several of them have been known only to a few who have had the good fortune to acquire them. For the most part illustrations were drawn from the actual cartridge. In a very few cases photographs and "miked" sketches were supplied by the owners.

Pen and ink drawings have been used as the medium for illustrations. Few techniques show up detail as clearly as simple line drawings, particularly when they are to be reproduced in the small actual size of cartridges.

It is my sincere wish that this pictorial digest of cartridges may serve to give you a clearer conception of this fascinating subject . . . and that it may help to place cartridges on an equal plane with their companion, the gun . . . as collectors' items.

Herschel C. Logan

Salina, Kansas

contents

	Page
The Cartridge Story (Frontispiece)	
Cartridge Development Chart	
Cartridge Chronology	1
Paper	11
Combustible	21
Separate Primed	29
Self-Contained	41
Patent Ignition	51
Rim Fire	59
Center Fire—Part 1	75
Center Fire—Part 2	103
Shot	157
Blank	173
Appendix	179
Primers	180
Bullets	183
List of Patents (to 1878)	187
Manufacturers Headstamps	189
Cartridges of the '90's	192
Comparative Dimensions	198
DWM Markings	199

CARTRIDGE DEVELOPMENT

Even though some of the later periods appear to have preceded earlier steps on the chart, it does not alter the chronological order. For instance the rim fire was invented in 1845 in France... yet it did not come into general usage here until nearly twelve years later. However, the earliest accepted date in each step is used in preference to the mere period of use.

Dates and periods generally accepted by authorities have been used in the compilation of this development chart. Since it is virtually impossible to correctly designate definite production dates, such a chart can only give a general idea of the origin and period of each development step. It will be observed that the twenty-year period from 1850 to 1870 contained all of the cartridge steps. It was perhaps the greatest development period in the history of ammunition.

CENTER FIRE
CENTER FIRE, *early*
RIM FIRE
PATENT IGNITION
SELF CONTAINED
SEPARATE PRIMED
COMBUSTIBLE
PAPER

Dates: 1800 1810 1820 1830 1840 1850 1860 1870 1880 1890 1900

THE TRUE ORIGIN OF THE PAPER CARTRIDGE
...like that of gunpowder, is veiled in the mists of the past.

x

CARTRIDGE CHRONOLOGY

Gathered from many sources here and there, it is believed that this is the first time such a compilation has been made available in this form. It is hoped that it will help to acquaint both laymen and collectors with the interesting story of cartridge development.

DATE	EVENT
A. D. 846 . . .	*Liber Ignium*, a manuscript by Marcus Graecus, which is in the National Library of Paris, describes an explosive compound . . . composed of six parts saltpetre, two parts sulphur and two parts charcoal. (Practically the same as the present day formula.)
1242 . . .	Friar Roger Bacon, in his essay, *De Mirabili Potestate Artes et Naturae* gave a formula for gunpowder, concealed in an anagram.
1313 . . .	Many authorities assign this date as the year in which Berthold Schwartz, a German Franciscan monk, discovered gunpowder.
1314 . . .	Town records of the city of Ghent, Flanders (Belgium), disclose that guns and powder were shipped to England.
1331 . . .	At the siege of Licante it is recorded that the Moors used gunpowder as a propellant for stone balls.
1346 . . .	At the battle of Crecy gunpowder was used as a projectile propellant.
1498 . . .	Barrels with straight grooves were used by citizens of Leipsic for target practice.
1520 . . .	Augustin Kutter perfected a so-called rose or star-grooved barrel having a spiral form.
1550 . . .	It seems likely that paper cartridges were developed during the matchlock period. It is believed that they were used by mounted troops in the second half of the 16th century. These early cartridges contained only the charge of powder to expedite loading while on horseback. Later the ball and powder were combined into one unit.
1560 . . .	P. Whitehorne wrote of "bagges of linen or paper" which contained the powder charge for cannon.

DATE	EVENT
1586 . . .	One of the cases used to carry small arm ball cartridges is in the Dresden Historical Museum. Partly filled with cartridges it is from the bodyguards of Christian I (1586-1591). The case held ten to fifteen cartridges.
1590 . . .	Sir John Smythe in his writings mentions this type (paper cartridges) as ". . . cartages with which musketeers charge their pieces both with powder and ball at one time."
1597 . . .	*Corona e palma militare* by Capo Bianco gives what is believed to be the first description of the complete paper cartridge. He wrote that it had long been in use by Neopolitan troops.
1600 . . .	Gustavus Adolphus (1594-1632) introduced the paper cartridge to his troops. Each man carried twelve in a leather case upon his back.
1663 . . .	*Pepy's Diary*, published in November of this year, describes the violent explosive qualities of fulminate of gold.
1742 . . .	*New Principles of Gunnery* by Benjamin Robins described his ballistic pendulum with which he measured the velocity of projectiles. Much of his experimental data is included in this early book.
1774 . . .	Louis XV's army physician, a Dr. Bayen, discovered the detonating properties of fulminate of mercury.
1788 . . .	Berthollet discovered that potassium chlorate when mixed with a substance would ignite when struck a blow.
1799 . . .	The composition of mercury fulminate was discovered by a man named E. C. Howard.
1802 . . .	The du Ponts began the manufacture of powder in America.
1807 . . .	A Scottish clergyman, Rev. Alexander Forsythe, obtained a patent for applying fulminate of mercury and other compositions to the ignition of gunpowder by detonation by a blow.
1812 . . .	M. Pauli (or Pauly) a Swiss working in Paris, developed a breech-loading gun which was chambered for a cartridge. The powder charge and ball were contained in a combustible envelope . . . which in turn was seated in a brass head not unlike the modern shotgun head. The primer consisted of a pellet placed in the primer pocket of this brass head . . . thus producing a self-igniting center fire cartridge.

DATE	EVENT
1814 . . .	Joshua Shaw, an artist of Philadelphia, invented the percussion cap. His first caps were of iron for repriming. In 1815 they were made of pewter and from 1816 on for fifty years they were made of copper.
1816 . . .	A copper tube containing detonating powder was patented by Joseph Manton of England.
	The U. S. Martial pistols were reduced in caliber from .69 to .54.
1817 . . .	The U. S. Government ordered 100 Hall breech-loading rifles. Breechloaders permitted the use of a ball larger than the bore . . . thus allowing less gas to escape than with the muzzle-loading system.
1820 . . .	A Frenchman by the name of Gevelot began the manufacture of fulminate of mercury for percussion forms of ignition.
1823 . . .	Dr. Samuel Guthrie, an American physician, developed a fulminate of such composition that it could be rolled into small pills or pellets. These were used for a number of years as the detonating agent for pill lock arms.
1825 . . .	Cartridges consisting of the powder charge in a paper case was the first ammunition used in the Hall breech-loading rifle. The end of the cartridge was bitten off by the soldier and the powder poured into the open chamber . . . the ball then being pressed in on top of the powder by the thumb. The paper case was then often used as a wad.
1826 . . .	A patent (No. 3,355) was issued in France to Galy-Cazalat for . . . "a case of leather or parchment. Priming in depression in base." This was one of the first self-contained cartridges known.
1827 . . .	Johann Nicholas Von Dreyse, a German gunsmith of Sommerda, began experiments with the needle-gun . . . forerunner of all modern bolt action rifles.
1828 . . .	Rifled firearms using a cylindrical elongated projectile developed by Delvigne were placed on trial at Vincennes.
1829 . . .	Pottet, a contemporary of Pauly secured a French patent, No. 3,930, for a centerfire cartridge with a removable base. It had the fulminate in a pocket or on a nipple.
1831 . . .	An English patent, No. 6,196, provided for a cartridge carrying a cap in its base.

DATE	EVENT
	Maj. Berner evolved a new system which it was hoped would unite the ease of loading of the smoothbore with the accuracy of the rifle. His gun was known as the "two-grooved Brunswick infantry or Oval rifle."
1832 ...	M. Braconnot of Nancy, France, discovered a pulverulent and combustible product (guncotton) while treating starch with a concentrated solution of azotic acid.
1836 ...	M. Le Faucheux of Paris is given credit for inventing the pin fire ... probably the first self-exploding cartridge to enjoy general use. His first cartridge was a cardboard and brass combination, not unlike the modern shotgun shell.
1837 ...	In England the Light Brigade was armed with a "two grooved rifle" ... which took a belted bullet.
1839 ...	Experiments in Sweden of the pointed bullet versus the round. Same type of experiments conducted in Switzerland and Saxony several years later.
1840 ...	The percussion form of ignition was adopted by the French government.
1841 ...	Hanson and Golden, two Englishmen, obtained a patent, No. 9,129 on a hollow base bullet containing fulminate. A peculiar shaped bullet called a "cylindro-conodical" ball having a hollow base and introduced by Capt. Delvigne of France was experimented with at Liege by a board of Russian and Belgian officers. Austrian Infantry replaced the flintlock with the new percussion primer lock as designed by General Augustin. A Zurich engineer by the name of Wild felt it impractical to drive the bullet into the grooves, as was usual with rifles, with enough force to injure its sphericity. He developed a rifle with a bore of from six to eight shallow grooves with a slight twist. The cartridge for his arm had its ball and patch attached.
1842 ...	Consideration given by the U. S. Ordnance Dept. for the converting of the Flintlock Muskets to percussion. The Model 1842 musket was the last regulation smoothbore, and the first regulation percussion.

DATE	EVENT
1844 ...	Previous to this date most of the military arms had used a round ball.
1845 ...	The U. S. Government began the manufacture of copper caps on a large scale ... developing machinery for quantity production.
	Dr. Edward Maynard, a dental surgeon of Washington, D. C., invented the tape primer ... a strip of paper having small amounts of detonating powder sealed in at regular intervals.
	Flobert of France is generally credited with the invention of the BB cap. It was used in the so-called "saloon pistol" ... which at that time was quite popular abroad. The cartridge consisted only of a copper shell containing fulminate and a small ball.
1846 ...	Adapting the use of guncotton to the loading of firearms was discovered by Schoenbein of Basle.
	The French Army adopted the solid cylindro-conodial bullet introduced by Captains Minie and Tamisier. Capt. Tamisier is the one who suggested using a number of cannelures on the bullet instead of the one circular groove.
	Col. Jacobs developed a four groove conical bullet for his four groove Jacobs rifle.
	M. Houllier, a French gunsmith, took out a patent, No. 1,936, in this year which included not only pinfire, but our present rimfire system and early centerfire system. However, most important was his employment of thin copper or brass as a complete case for a cartridge ... regardless of the type of primer used.
	Capt. Minie developed a bullet with a sheet-iron wedge, or culot within the base.
1848 ...	W. Hunt secured a patent, No. 5,701, for a lead projectile containing powder in its base, and covered by a metal cap.
	The Dreyse needle gun invented by Von Dreyse and officially adopted by the Prussian Army in 1840 was first issued to troops in this year.
	Bavaria adopted the Thouvenin system. This system, invented by Col. Thouvenin of the French Artillery, consisted of using

DATE	EVENT
	a pin or stem projecting into the barrel from the breech plug. This permitted the ball to be rammed without disturbing the powder which was poured around the pin. The pin was long enough to project slightly above the powder. The pin or "tige" also served as an anvil, forcing the ball, when rammed, into the grooves and causing them to be filled.
1849 . . .	Saxony adopted the Thouvenin system and altered the Jager rifles to it.
1850 . . .	Both the Lancaster and Needham shotgun cartridges were in general use during the early fifties.
1851 . . .	The U. S. Army purchased 393,304 paper cartridges.
1852 . . .	Beginning of experiments on the adaptability of Capt. Minié's elongated bullet for our armed service. The bullet, minus the wedge, was later adopted.
	Christian Sharps patented his disk primer. His patent embodied the use of a disk primer fed from a tube located in the lock plate.
	Similar to the combustible envelope cartridge, except that it was made of linen instead of paper . . . the Sharps linen cartridge was a step forward. The linen case would stand more abuse than the fragile paper variety.
1854 . . .	A bullet with a primer in its base was patented (No. 10,698) by Gaupillat of France.
	Conversion of the 1822 and 1842 muskets to .58 caliber rifle using the conical bullets.
	The firm of Schuyler, Hartley & Graham, early noted arms and ammunition dealers, opened for business in March.
	The first really successful metallic cartridge, self contained and reasonably moisture proof was patented by Smith & Wesson. It was first produced for revolvers late in 1857.
1855 . . .	Beginning of a ten-year period which saw more different types of breech-loading arms developed than any other equal period in the history of arms. Many of these arms required their own peculiar type of cartridge.
	Preparations were made by the government to develop a new

DATE	EVENT
	model rifle to incorporate the Maynard primer . . . and to standardize on a rifled barrel and a conical bullet. Caliber of all long arms was changed from .69 to .58 caliber.
	During this year Colt and Eley Brothers collaborated in the improving of paper cartridges.
1856 . . .	Geo. W. Morse invented a soft metal cartridge case with a priming apparatus to seal the breech and vent of a breech-loading gun at firing.
	The cartridge used in the Volcanic arms was patented by Smith & Wesson. Its lead bullet contained powder and a primer containing fulminate.
1857 . . .	A cardboard tube cartridge having the bullet at the rear of the powder charge was patented by J. D. Greene.
1858 . . .	The patent of the Morse metallic cartridge with its inside anvil and perforated disk containing a percussion cap marked the first really important step forward in the history of metallic centerfire cartridges.
1859 . . .	Edward Maynard secured a patent for a separate primed metallic cartridge.
	Willard C. Ellis and John N. White were granted a patent for a front loading, cup-primed cartridge.
1860 . . .	The Burnside cartridge with a grease chamber around the bullet was patented by Geo. Foster. It was one of the first brass case cartridges produced in quantity in this country. It is believed to have the first inside lubrication of a bullet.
	A metallic cartridge having a side lip for fulminate was patented by E. Allen.
1864 . . .	D. Williamson secured a patent on a cartridge having the fulminate in a teat at the base of the metal case . . . now referred to as "Teat Cartridge."
1865 . . .	The Martin, first metallic centerfire cartridge, adopted for the service, was used in the 1865 Allen alteration of the muzzle-loading muskets.
	Ethan Allen secured a patent (No. 47,688) for a brass reloadable cartridge case using a rimmed percussion cap (not unlike a tiny rimfire cartridge) in the head.

| DATE | EVENT |

| | A .58 cal. inside-primed cartridge was used in the second Allen alteration of the Springfield musket. |

The Crispin cartridge with the fulminate in an annular rim surrounding the metal case was patented by S. Crispin.

(There was also a centerfire Crispin with a primer in the base. The annular rim on it was utilized to hold the cartridge in place in the chamber).

1866 . . . The U. S. Government began extensive experiments with centerfire types of cartridges.

New Haven Arms Company was reorganized into the Winchester Repeating Arms Company. Rimfire cartridges manufactured by this firm have, since their founding, carried the letter "H" as a head stamp . . . in honor of B. Tyler Henry, their chief engineer.

In March of this year Hiram Berdan secured a patent (No. 53,388) which provided for . . . "the fixed permanent teat-like projection at the bottom of the cup in the head of a cartridge, in combination with the pellet or other priming inserted into and protected within the said cup."

In September of this year the English service officially adopted the Snider converted Enfield and the Boxer cartridge as their standard arm and ammunition. This arm was the first breechloader of the English Army.

An auxiliary chamber provided with a nipple for a percussion cap was patented by D. Williamson for his combination cartridge and percussion derringer.

1867 . . . In August of this year the Union Metallic Cartridge Company opened for business. It was founded by Marcellus Hartley and associates (Schuyler & Graham).

Schultze powder with a nitrocellulose base of nitrated wood fibre was developed. It was suitable only for sporting purposes.

1868 . . . Development of a cartridge for the Thuer's alteration of the Colt revolver. It was first made with an inside primer, then later with an outside primer.

Col. S. V. Benet, of the Frankford Arsenal, introduced an experimental solid-head cartridge . . . which also contained an inside cup anvil.

DATE	EVENT
	Gen. Benjamin F. Butler and associates organized the United States Cartridge Company at Lowell, Mass.
1869 ...	A. C. Hobbs was awarded a patent (No. 94,743) for a primer. His claim, "a percussion for guns enclosed between varnished surfaces."
	John A. Roebling, a noted American engineer, writing in the *Scientific American*, said ... "It may well be questioned whether any invention in the arts of gunnery since the introduction of gunpowder, was a longer stride in advance than the invention of the copper percussion cap."
1871 ...	Martin's folded-head cartridge, in which the primer pocket was formed out of the continuous metal of the case, was produced in considerable quantity during this year.
	The .577 Snider Boxer cartridge was dropped by the English Army in favor of the then new .577-450 Martini-Henry cartridge, which was first made in a Boxer-type case and later in the regular drawn case.
	The .577-450 Martini-Henry was one of the first English attempts to bottleneck a cartridge case.
1873 ...	The U. S. Army selected a breech-loading rifle based on the Allen system ... and the caliber was reduced to .45.
1880 ...	Majors Bode and Rubin, working jointly in Switzerland, developed a metal jacket bullet.
1881 ...	This year saw the renewable or crimp type primer perfected to the state where it was to be used on all future cartridges manufactured for the U. S. services.
1883 ...	Gevelot, the French ammunition manufacturer, bought out Gaupillat to form "Gevelot et Gaupillat." A year later the firm name was changed to "Societe Francaise des Munitions."
1885 ...	A French chemist by the name of Vieille developed the first smokeless powder to prove satisfactory. This powder was first used in the Model 1886 Lebel rifle ... noted as the first rifle of a world power of so small a caliber, 8mm.
1887 ...	Bullard's catalog lists the .50-115 Express cartridge. This was one of the first, if not the first, semi-rimless cartridges on the American market. It was also the first solid drawn head with-

DATE	EVENT
	out a balloon primer pocket projecting into the interior of the case.
	Peters Cartridge Company organized at Kings Mills, Ohio.
1888 . . .	A process termed "colloiding," in which guncotton was made to burn and not explode, was worked out by the Frenchman Vieille.
	Ballistite, a double based powder containing both nitro-cellulose and nitroglycerine was patented by Alfred Nobel.
1892 . . .	The Model 1892, .30 Krag Jorgensen, was the first U. S. rifle to use smokeless powder and a reduced caliber. The bullet had a lead core encased with a cupro-nickeled jacket . . . blunt nose and flat base.
1893 . . .	The Borchardt, first automatic type handgun, was designed and chambered for ammunition which has, more or less, been a model for all automatic ammunition since.
	Francis G. du Pont developed a hard grain smokeless powder with a fairly high nitrogen content.
1898 . . .	Development of the "H-48," a non-mercuric fulminate for primer fillers, by the Frankford Arsenal.
1900 . . .	Peters Cartridge Company produced the first crimped .22 long rifle cartridge. It was known at that time as the Smith & Wesson Long.
1902 . . .	Remington Arms Company merged with the Union Metallic Cartridge Company to become familiarly known as Remington-UMC.
1906 . . .	The U. S. service adopted the German type "spitzer" bullet which had a pointed nose and a flat base . . . weight 150 grains.
1924 . . .	The 172-grain gilding metal encased bullet having a nine degree chamfer at the base was adopted by the U. S. It is commonly known as "boattail."
1934 . . .	E. I. du Pont de Nemours & Company bought out Remington-UMC.
	Peters Cartridge Company was absorbed by du Pont and Remington.

paper

BALL

BUCK AND BALL

BUCKSHOT

PAPER

The first cartridges were very simple . . . a powder charge wrapped in paper and tied with a string. For the most part they were used only in an emergency and to expedite loading while on horseback. The ball was a separate unit to be inserted by itself.

Later, someone conceived the idea of combining both powder and ball into a paper case, a development that was to carry on for nearly three centuries.

Such cartridges were used in matchlock, wheel lock, flintlock and percussion arms of various types. Oftentimes they were greased or oiled, not only to lubricate the barrel, but to protect the powder charge from dampness.

Paper cartridges used both round balls and conical bullets for projectiles. Those with only one large ball or bullet were referred to simply as .69 Ball, .69 Pistol, Musketoon or Musket. Those with one large ball and three smaller ones were called .69 Buck and Ball, while those with twelve small balls were labeled .69 Buckshot, or whatever the caliber happened to be in each case. All were used in smoothbore arms.

The original instructions directed the soldier to tear the cartridge open, pour the powder into the barrel, then unwrap the bullet and ram it down. Contrary to this, they soon developed their own loading technique. This many times merely included the biting off of the end, pouring the powder charge, less a small amount which was used for the priming pan, into the barrel and then ramming down the ball. Paper and string were then rammed down on top to serve as a wad.

PAPER CARTRIDGE
METHOD OF MANUFACTURE

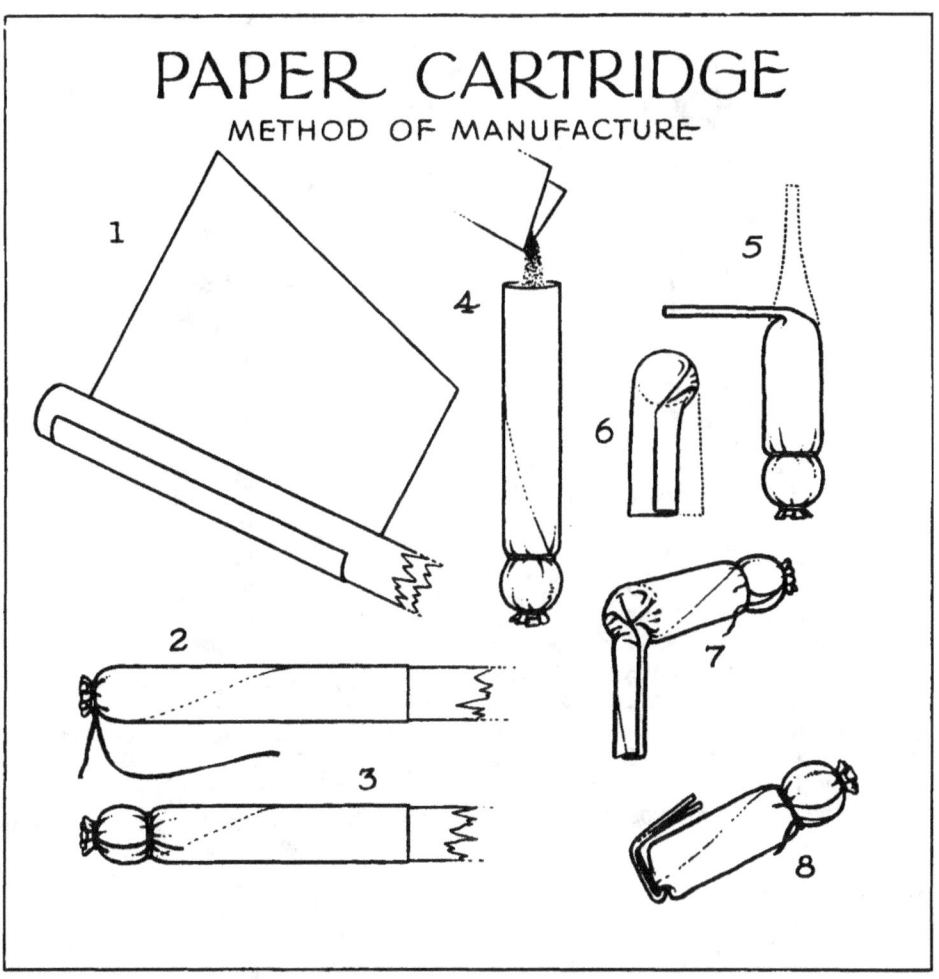

1. Cartridge paper 4 1/3 inches wide by 5 1/4 inches in overall length, (for .69 caliber) wrapped around a cartridge stick of .65 diameter.
2. The stick slipped back from end allowing paper case to be "choked" or tied.
3. Stick is removed and ball is inserted, followed by stick which holds case in shape while ball is being secured by two half hitches.
4. Fifty grains of black powder is poured into case.
5. End of case is pinched and bent over at right angle.
6. Right side is folded in to center.
7. Left side is folded in after which the "tail" is folded up and over to form . . .
8. The completed cartridge.

CALIBER 36

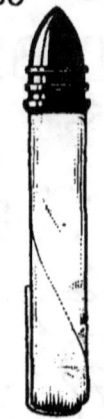

(Sharps 90 bore Paper)
Total length 2⅜ in.
1¹³⁄₁₆ in. buff colored paper case
Conical lead bullet

These 90 bore (.36 cal.) cartridges were used in the Sharps slanting breech sporting rifles. The slanting breech rifle was patented by Christian Sharps in 1848. Sharps Pat. disc priming magazine is to be found on the early issues of this rifle. Later the Lawrence Pat. priming magazine replaced the earlier Sharps method.

CALIBER 44

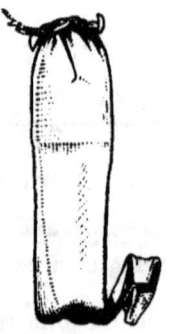

(44 Army)
Total length 1¹³⁄₁₆ in.
Paper case is lemon color for easy identification.
216 gr. lead, conical bullet

25 grs. black powder

These U. S. (1858) paper cartridges were used in the Army Dragoon revolvers. (See Plate 160 *Hand Cannon to Automatic*)

CALIBER 52

(32 Bore Sharps Paper)
Total length 2⅝ in.
1¹¹⁄₁₆ in. paper case, tied with cord
Conical lead bullet

This 32 bore cartridge was later called the .52 caliber. It was used in the Model 1859 Sharps B. L. percussion military rifles and carbines. Later the .52 caliber size was made by cementing the paper tube to the base ring of the bullet. Still later this size developed into the familiarly known Sharps linen type.

CALIBER 54

(54 Martial Pistol)
Total length 1½ in.
Cream colored paper case
½ oz. spherical lead ball

Paper cartridges such as this were used in both flint and percussion martial arms from 1815 to around 1860. An interesting note on the one illustrated is that it came from an old box, hand labeled . . . ".525 Colt Musket." However, its measurements are identical with the .54 caliber cartridge, whose ball does mike .525. Due to the fouling of the barrels by the old black powder, the balls of most of the early cartridges were made smaller than the actual bore measurement . . . hence a .525 ball for a .54 caliber. Ammunition such as this was used in the Black Hawk, Seminole and Mexican Wars.

CALIBER

(Unknown)
Total length 2 1/32 in.
1⅛ in. paper case
Conical lead bullet of .650 diameter

It is said that this large cartridge will fit the chambers of the .56 Colt Revolving Musket. However, it is larger than the combustible cartridge which was made by the Colt Cartridge Works for the revolving musket. It is illustrated here as another type of these early cartridges.

CALIBER 58

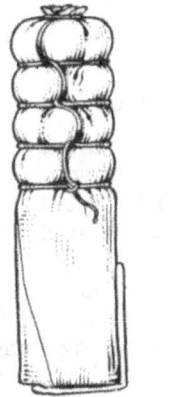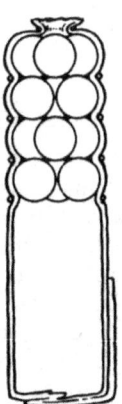

(58 Buckshot)
Total length 2 5/16 in.
Buff colored paper case
12 lead balls weighing 43 grs.
75 grs. black powder
Total weight 625 grs.

Used in the 1855 Musket of .58 caliber, these buckshot cartridges must have been the granddaddy of the present day riot gun charge. This caliber was also made in the "buck and ball" load—one large ball and three buckshot.

CALIBER 58

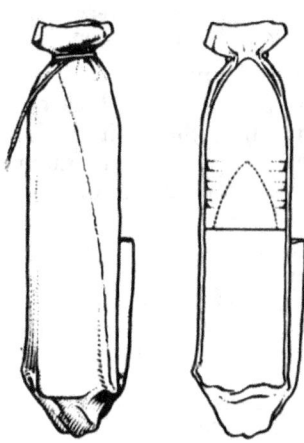

15

(58 Springfield)
Total length 2½ in.
Buff colored paper case
500 gr. conical lead ball
  (Minié Ball) with concave base
60 grs. black powder

Instead of using an iron cup in the base of the Minié ball, as was invented by Capt. Minié, to spread the bullet to fit the rifling upon discharge of the propellant . . . the army found that the powder charge itself was sufficient to expand the deep, hollow-based bullet. This cartridge was developed for the 1855 Springfield musket.

CALIBER 64

(64 Paper)
Total length 1¹¹⁄₁₆ in.
White paper case, tied with white and red striped cord
393 gr. spherical lead ball
74 grs. black powder

This is the type of ammunition the soldiers under Gen. Jackson were using in their .69 caliber Hall carbines and rifles in 1818 during the Seminole Campaign in Florida.

CALIBER 69

(69 Pistol)
Total length 1¹³⁄₁₆ in.
Buff colored paper case tied with red and white striped cord
1 oz. spherical lead ball
50 grs. black powder

On March 9, 1799, the U. S. Government made its first contract with Simon North for 500 horse pistols of .69 caliber. Pistols of this caliber were used in the service until 1815. It was the .69 flintlock pistol that was the hand gun guardian of the then infant United States. (See plates 50, 54, 56 and 59 *Hand Cannon to Automatic*)

CALIBER 69

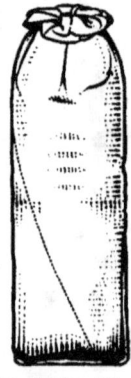

(69 Rifle Musketoon)
Total length 2⅛ in.
Buff, coarse grain, paper case
625 gr. conical lead bullet
50 grs. black powder
Made at Frankford Arsenal, 1859

The rifle musketoon was a short musket with a large smooth bore. Musketoons using this ammunition were first brought out in 1842. They were used by both the artillery and cavalry . . . as well as the sappers (engineers).

Label . . .
    RIFLE MUSKETOON, Cal: .69
        50 Grains M. Powder
        Weight of Ball 625 Grains
    FRANKFORD ARSENAL
            1859

CALIBER 69

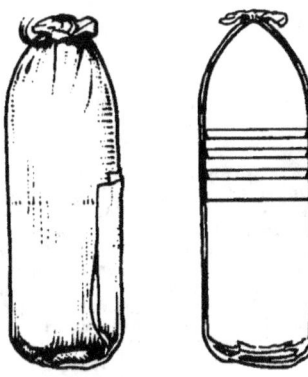

(69 Musket)
Total length 2 1/16 in.
Buff colored paper tied with light colored cord

Since the regulation smooth bore musket of .69 caliber gave way to the .58 caliber musket with a rifled barrel using a conical bullet in 1855, the life of this cartridge must have been very brief. It is known that some of the 1842 muskets and musketoons were rifled to take the new conical bullets, as a sort of experimental prelude to the actual reduction of caliber to .58.

CALIBER 69

(69 Musket)
Total length 2 5/8 in.
Heavy, coarse, buff colored paper case
1 oz. spherical lead ball

Apparently this item is one of the early experimental cartridges. It will be noted that the ball is partially exposed. It was used no doubt in the .69 caliber smoothbore muskets.

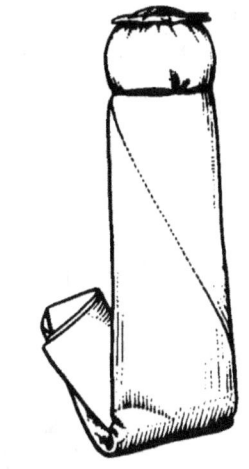

CALIBER 69

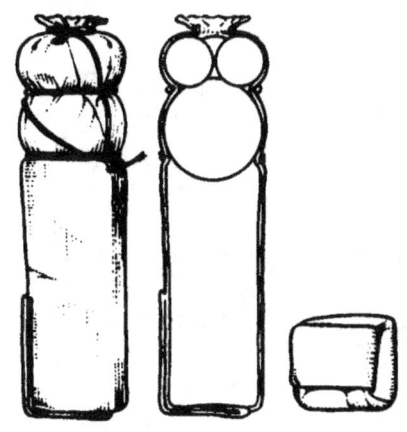

(69 Buck and Ball)
Total length 2 3/8 in.
Dark buff colored paper case tied with light brown cord
Three 43-gr. lead buckshot
1 433-gr. spherical lead ball
75 grs. black powder
Total weight 662 grs.

Another of the multi-ball loads for the .69 muskets and musketoons is this "Buck and Ball." The cross-section view shows how the three buckshots were placed on top of the large ball within the paper case.

The following descriptions of the Paper Cartridges on the opposite page are taken almost verbatim from the book . . .

A Brief Description
of the
MODERN SYSTEM OF SMALL ARMS
As Adopted in
The Various European Armies
By J. Schön

(Translated from the German by J. Gorgas, Capt. of Ordnance, U. S. Army, Dresden, 1855)

1 This cartridge, used in the Norwegian Breech-loading Rifle, is commonly of writing paper, and wound twice. One end laps over the groove of the bullet, and is tied with a thick greased woolen thread, and the other end is turned down over the charge . . . which is 70 grains. The bullet is 1.137 inch long, and 0.696 inch diameter. Weight 787 grains.

2 A Belgian cartridge which is wound double and pasted its whole length. The open end is drawn over the ball to the first groove to which it is tied with a woolen thread. The loose end of the cartridge is twisted and turned down over the end. The bullet is 1.125 inch long . . . .672 inch in diameter and its weight is 731 grains.

3 The cylinder of this Bavarian cartridge is of paper, unpasted and contains 58 grains of powder . . . from which the bullet is separated by a tie. In loading, the end folded down is not bitten off but opened, the powder poured in, the cartridge turned, and the bullet shoved out of it with the rammer. It is pushed home and rammed three times, which drives the tige from 0.15 inch to 0.09 inch deep into the lead, and fills the grooves. The bullet's length is 0.957 and its weight is 478 grains.

4 Used in a Prussian fortress wall piece (defensions-gewehr) this cartridge in which the bullet is placed, point inward, is tied between the powder and the ball, and both ends are turned down and not pasted. Powder weight 102 grains. The bullet is 1.23 inch long . . . .694 inch in diameter and weighs 777 grains.

5 This Belgian cartridge is made of thin cardboard rolled around once, and, for greater security and the preservation of the powder, is wrapped about with an envelope of common writing paper pasted, the end of which is inserted into the cylinder in such a way as to envelope the point of the bullet and separate it from the powder. The whole is then surrounded by a third wrapper unpasted and folded close over the base of the bullet . . . the other end being left long, is twisted and turned down. The cartridge is dipped in tallow as far as the cylindrical part of the bullet extends. The method of loading is as follows: The cartridge is bitten off above the powder, the powder poured in, the cylindrical part of the bullet inserted, the empty cartridge torn off, and the bullet shoved down without ramming on the charge, which is 85 grains of common musket powder. In loading in this way, without freeing the bullet of the paper, there will be no danger of the bullet moving forward when the gun is carried with its muzzle depressed. The bullet is 1.223 inch in length . . . .673 inch in diameter, and weighs 721 grains.

6 The cartridge cylinder of this Austrian item is not pasted, but folded down over the base of the bullet. The bullet is turned point inwards, and tied with a thread over the point, which prevents the access of the powder and gives the cylinder the required stiffness. The loose end is folded down and the primer attached to it. The bullet is 1.035 inch in length . . . .692 inch in diameter, and weighs 652 grains.

7 Hanoverian troops used this cartridge, which was made of fine but strong vegetable paper, not pasted. The loose end was folded twice at right angles and turned down in the Prussian manner. The bullet was 0.8 inch in length . . . .61 inch in diameter, and weighed 444 grains.

8 The cylinder of this Russian cartridge is double, has its edges pasted and is folded down over the base of the bullet, which has its point turned inward. It is tied in the groove with a woolen thread dipped in tallow. The other end is pinched and folded down. Another tie is made about a third way down the conical part of the ball, the object of which is to prevent the powder from shaking down. The bullet is 1.191 inch in length . . . 0.70 inch in diameter, and weighs 689 grains.

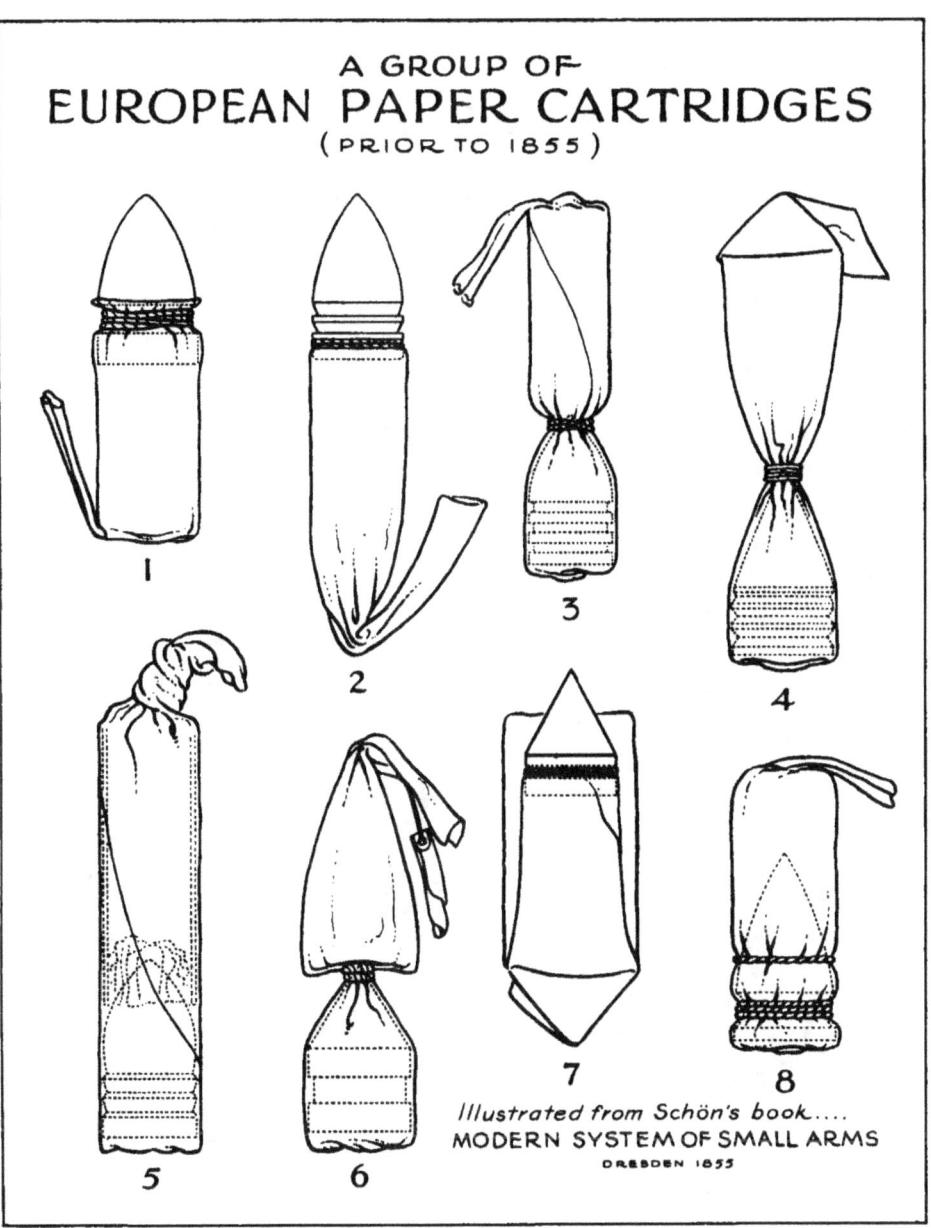

combustible

PAPER

LINEN

SKIN

COLLODION

COMBUSTIBLE

Two developments came in the first half of the nineteenth century. One was the conical bullet, the other the combustible envelope used in the manufacture of cartridges. Together with the newly invented percussion cap, they produced an advanced step in cartridge evolution.

Various materials, paper, linen, collodion, skin, etc., were used in forming the case for the powder charge. All were highly nitrated making them readily inflammable. It mattered very little whether or not the case broke during the loading since it would ignite anyway from the fire of the percussion cap.

It is regrettable that no record seems to be available as to who produced the first combustible cartridge. They were produced by several firms during the middle of the nineteenth century. Used both in hand guns and long arms they were, for the most part, put up in wooden hollowed containers or blocks. These containers, holding five or six cartridges, were drilled out so that each cartridge was in a separate compartment by itself. The wooden blocks were wrapped in oiled waterproofed paper upon which the label was printed.

The Sharps linen, to mention one outstanding example, was a product of this period of development. It was less fragile than the paper variety and became quite popular, along with the Sharps breech-loading rifle.

MAKING A COMBUSTIBLE CARTRIDGE

1. Cartridge paper, or linen cloth, cut in a strip of one and three-eighths inches (for the Army size .52 caliber Sharps) and of sufficient length to wrap twice around a cartridge stick.
2. Gluing the edge and securely fastening it forms the paper cylinder.
3. Cut a piece of thin bank-note paper, or gauze, three-fourths of an inch square.
4. Place this square of paper on end of cartridge stick... apply glue either to this paper or to the inside rear edge of paper case.
5. Insert stick with paper into case forcing it to the rear... and withdraw stick.
6. Allow case to dry after which charge with 60 grains of black powder.
7. Apply an adhesive to ball before seating it in case. Choke paper into grooves of ball to form...
8. The complete cartridge.

CALIBER 28

(28 Combustible)
Total length 1 3/10 in.
3/4 in. paper case
59 gr. conical lead bullet
Total weight of cartridge 67 grs.

The bullet in this cartridge is identical with those cast by the .28 caliber Colt mold . . . used in the early percussion pistols and revolvers.

CALIBER 31

(31 Paper Combustible)
Total length 1 1/8 in.
3/4 in. paper case
Conical lead bullet

Made for Colt, Whitney or Remington revolvers of 31-100 caliber. A cartridge of this type was patented by J. H. Ferguson, June 28, 1859, Pat. No. 24,548. The paper case was prepared according to the patent papers as follows: "First, the paper is treated with a compound consisting of eighteen parts by weight of nitrate of potassa, pure, and seventeen parts of sulphuric acid, pure, after which it is washed to free it from the soluble salts and excess of acid, and then dried between sheets of blotting paper. This preparation renders the paper highly inflammable, but not at all explosive . . . in order to render it perfectly waterproof, a light coat of shellac varnish is applied to one side only." A contemporary ad, in describing these new cartridges, says:

Combustible Envelope Cartridges

For Colt's, Remington's, Whitney's, Bacon's and all other revolvers using caps. They are put up in waterproof cases, are more convenient, and surer fire than loose powder and ball, and a revolver can remain loaded much longer, without injury, when charged with these cartridges instead of loose powder and ball.

CALIBER 36

(36 Colt Combustible)
Total length 1 1/4 in.
3/4 in. paper case
Conical lead bullet with two grease grooves

These cartridges were packed in a flat wooden container having a separate hole for each of the six charges. The wood box was wrapped in an oiled paper upon which was printed the following:

6 Combustible Envelope
Cartridges
Made of HAZARD'S Powder
Expressly for
Col. Colt's Patent
REVOLVING BELT PISTOL
address
Colt's Cartridge Works
Hartford, Conn.
U. S. America

CALIBER 36

(36 Skin Cartridge)
Total length 1 5/16 in.
Cylindrical . . . round nose . . .
 lead bullet

An English cartridge illustrated in its paper carrying case. The strip of paper to the left of the bullet is used to quickly remove the case before the combustible skin cartridge was inserted into the chamber of the gun. On the blue label around the case is the following wording:

Capt. M. Hayes, R. N.
Patent
SKIN CARTRIDGES
Manufactured by
BROUX & MOLL   LONDON

CALIBER 38

(38 Eley Colt)
Total length 1 5/16 in.
7/8 in. skin case, wrapped with
 small cord
Conical lead bullet

Shown beside the cartridge is its paper carrying case. The black ribbon is used in removing the case from the cartridge before inserting it into the chamber of the gun. Eley Bros. were one of the most noted of English cartridge manufacturers. The cartridge pictured here was their bid for the ammunition business of owners of .38 caliber Colt and other percussion revolvers.

CALIBER 44

(44 Bartholow Waterproof)
Total length 1 5/16 in.
1/2 in. collodion coated case
Conical lead bullet

On May 21, 1861, R. Bartholow secured a patent for a waterproof combustible cartridge which was described as follows in Patent No. 32,345 . . . "Nitrate of potassa, charcoal, sulphur and chlorate of potassa mixed with shellac, pressed into form of cartridge, and coated with collodion." Cartridges for the Army were packaged in containers labeled "for Army Holster Pistol." Those for Navy use were labeled "Navy Pistol Size." (See plate 160 *Hand Cannon to Automatic*)

CALIBER 44

(44 Colt)
Total length 1 5/16 in.

¾ in. paper case
Conical lead bullet

One of a series of cartridges used in the famous .44 Colts of Civil War and frontier days. This one is described by its original label as follows:

6 Combustible Envelope
Cartridges
Made of HAZARD'S Powder
Expressly for
Col. Colt's Patent
REVOLVING HOLSTER PISTOL
Address
Colt's Cartridge Works
Hartford, Conn.
U. S. America

CALIBER 44

(44 Colt Skin)
Total length 1$\frac{11}{10}$ in.
1⅛ in. skin case
Conical lead bullet

While the cleaned and chemically treated pigs' intestines were still wet they were stretched over forms of the same shape as the cartridge case. After the skin cases were dry and the powder and bullet had been put in place the outside of the case was given a coating of gutta-percha varnish. For the lubrication in the grooves around the bullet the following was used: ... three parts best tallow, two parts wax (vegetable preferred), and one part native gutta-percha, melted together in the order named; and the scum removed while near the boiling point. This cartridge Patent No. 33,611 was patented on October 29, 1861, by Wm. Mont Storm.

CALIBER 44

(44 Colt, Hazard's Waterproof)
Total length 1¼ in.
Manufactured by the Hazard Powder Co., Hazardville, Conn.

R. O. Doremus and B. L. Budd patented this waterproof cartridge, Patent No. 34,725. The powder is formed into shape upon a pin projecting from the base of the bullet ... or is pressed into the base of the concave bullet if no pin is used. The powder may be made up of layers of varying combustibility in order to provide greater acceleration. It is interesting to note one paragraph in the patent papers concerning this cartridge. It says: ... "Among the important features of this ammunition — besides possessing the qualities named in our other application as due to powder in the solidified form—are readiness to use, while in consequence of the absence of paper covering, the men need no longer be condemned for infantry who have lost their side teeth." In other words with this new patent they no longer had to bite off the end of the cartridge. The box label for this cartridge reads as follows:

Pressed Waterproof
CARTRIDGES
For Colt's Army Pistol
Patented March 18, 1862
Manufactured by the
HAZARD POWDER CO.
Hazardville, Conn.

CALIBER 44

(44 Skin Cartridge)
Total length 1⅛ in.
Round nose, lead bullet

Another of the Capt. M. Hayes patent combustible cartridges, pictured in its paper carrying case.

CALIBER 46

(46 for Colt)
Total length 1⅝ in.
Conical lead bullet

Algernon K. Johnston of Middletown, Conn., and Lorenzo Dow of Topeka, Kansas, on October 1, 1861, secured a patent (No. 33,393) for a waterproof cartridge. On January 7, 1862, they secured a similar patent (No. 34,061). A paragraph from the first patent reads: "Then, in order to render the cartridge compact and waterproof, or highly impervious to moisture, we coat the cartridge so constructed with collodion, or with a film of any substance obtained by dissolving cotton, flax or other textile matter in the ether of some acid, either alone or in combination with alcohol." The following label would indicate that Johnston & Dow were patentees rather than manufacturers:

Johnston's & Dow's
WATERPROOF & COMBUSTIBLE
CARTRIDGES
Manufactured by
Elam O. Potter, New York, U. S.
for Colt's Army Pistol 46-100
with powder made expressly for
these cartridges by
HAZARD POWDER CO.

It is not unusual to find early cartridges marked as being larger than the usual bore of the arm to which they belong. A ball of larger diameter than the bore was used, so that it would take the rifling. Hence, this .46 caliber would be used in the regular .44 Colt army revolver.

CALIBER 52

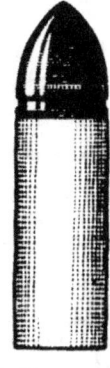 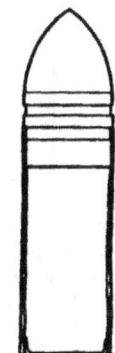

(52 Sharps Linen)
Total length 2¹⁄₁₀ in.
1⅜ in. linen case with nitrated paper base
Conical lead bullet

Manufactured by:
Sharp's Rifle Mfg. Co.
Washington Arsenal
A. G. Fay, Potter & Tolman
H. W. Mason and others

This is one of the most famous of all of the combustible types. It was used in the Sharps Sporting and Military Arms before and during the Civil War. Every student of arms history is familiar with the Sharps "Beecher's Bibles." These were the rifles sent into Kansas in 1857 in boxes labeled

"Beecher's Bibles." When John Brown was captured at Harpers Ferry in 1859, 102 Sharps carbines were taken from him. True it is that the .52 Sharps linen is one of the cartridges which not only saw history in the making, but helped to make it. There were 16,306,508 Sharps cartridges purchased by the government during the Civil War. These cartridges were listed as late as 1897 in the Hartley & Graham catalog.

CALIBER 54

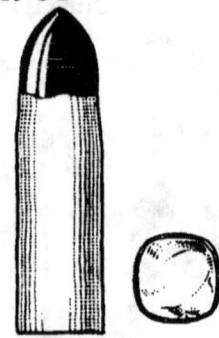

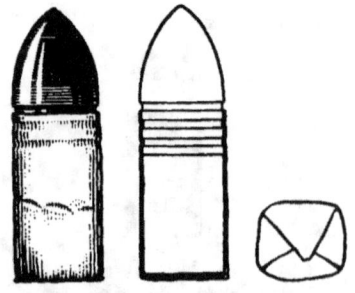

(54 Starr Linen)
Total length 1 13/16 in.
1 7/16 in. linen case with nitrated paper base
Conical lead bullet
80 grs. black powder
Manufactured by Johnston & Dow and others

Similar to the Sharps linen, the .54 Starr was developed for the Starr percussion carbine. A considerable quantity of these arms, (20,601) were delivered to the U. S. War Department during the period between July 30, 1863, and Aug. 20, 1864. A total of 25,603 were used during the Civil War. For these the government purchased 6,860,000 cartridges.

CALIBER 56

(56 Merrill)
Total length 1 3/8 in.
1 in. paper case
Conical bullet
40 grs. black powder

This is the scarce size of the Merrill combustible cartridge. Neither Satterlee nor Sawyer mention this caliber. The label from the original box reads as follows:

10 cartridges for
MERRILL'S PATENT CARBINE
Cal. 56-100 - Gr. 40
Address
The Merrill Patent Fire Arm Manufacturing Company
Baltimore, Md.

CALIBER 56

(56 Colt Revolving Rifle)
Total length 1 7/8 in.
1 1/10 in. paper case
Conical bullet of .583 diameter
Manufactured by Colt Cartridge Works

This cartridge was used in the Model 1855, .56 Colt revolving rifle or carbine. These arms were the 5-shot cylinder models... used by the militia in some states. A few came into the Civil War when the militia companies were mustered in. They were not too popular with the troops, however, due to the loud report and flash so close to the face and also to the fact that occasionally several chambers went off at once.

separate primed

PAPER

CLOTH

RUBBER

METALLIC

SEPARATE PRIMED

The invention of the percussion cap precipitated a period of experimentation, which it was hoped would lead to quicker loading and firing.

The paper, linen and other materials of the combustible era gave way to metallic cases, even though there were a number of cases of paper, and paper and foil, in the early part of this separate primed period. These separate primed cartridges were ignited from the fire of an ordinary percussion cap, disk primer or tape primer.

Two cardboard cased cartridges should be mentioned here. One was the Whitworth tube used in the Whitworth rifle. It has a long, round nose bullet, seated flush with the top of the tube. The rear of the tube was torn off exposing the powder charge when loading into the barrel. The other, the Greene, was the more unusual. It had a similar tube but in this case the bullet was behind the powder charge and served as a gas seal.

First of the metallics of this period was the well known Maynard and Burnside brass cartridges. While they were both separate primed, there the similarity ceased. In appearance they were as unlike as two cartridges can be. The Maynard was a tubular brass case with a wide flat base soldered to it. In the center was a small pin hole through which the fire from the percussion cap (or tape primer) passed.

The Burnside on the other hand had a tapering case not unlike an ice cream cone. Around the bullet was a heavy bulge which was used as a grease chamber. This cartridge, too, had a small pin hole in its inverted base.

The following pages will picture several other types developed during this period.

CALIBER 35

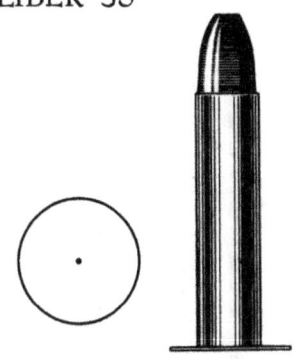

(35 Maynard Percussion)
Total length 2 in.
1$^{17}/_{32}$ in. brass case, .400 dia.
Conical, flat nose bullet

Used in the Maynard sporting rifles, these cartridges were among the first, if not the first, reloadable brass cartridges. They were used largely in the Model 1865 arm. The Maynard cartridge was patented by Edward Maynard first on June 17, 1856. This was for a metal shell with the rear aperture closed by waxed paper. On January 11, 1859, he secured another patent (No. 22,565) which provided for, "a perforated steel disk soldered to perforated base of brass shell." This is the type of cartridge illustrated, except the steel disk gave way to brass. The small hole in the head admitted the flame from the percussion tape or cap to ignite the powder within the case. The label from the box containing these interesting cartridges reads thus:

10
METALLIC CARTRIDGES
(loaded)
35-100 Calibre
EDWARD MAYNARD, Patentee
June 17th, 1856   January 11, 1859
Re-issue May 28, 1861
Manufactured by
MASSACHUSETTS ARMS CO.
Chicopee Falls, Mass.

CALIBER 36

(36 Marston)
Total length $^{7}/_{8}$ in.
1$^{7}/_{32}$ in. light green paper case
.330 dia.
Conical lead bullet .350 dia.

This rare cartridge was patented May 18, 1852, (No. 8,956) by Marston and Goodell. The shell was to be of paper or metal with a leather or paper base. It was used in the Marston single shot pistol. This arm was described in the booklet entitled *Art and Industry at the Crystal Palace in 1853*. This was an important industrial exposition of that day. It is said that the edges of the leather head were greased. When the cartridge was fired, the disk remained, to be pushed out by the bullet of the next cartridge. The disk being greased, this operation kept the barrel in clean condition. A patent, (No. 8,273) was also secured in France in 1852. (See plate 88, *Hand Cannon to Automatic*)

CALIBER 36

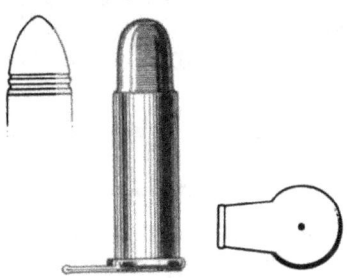

(36 Sharps "Mule Ear, Lop Ear or Flop Ear")
Total length 1½ in.
1$^{1}/_{16}$ in. brass case with projection on base

Dia. at head .414
Dia. at mouth .409
Round nose, lead bullet

These unique cartridges were produced in 1857 for Sharps pistol rifles, produced in Sharps Philadelphia factory. A shallow mortise was milled into the upper side of the breech to accommodate the projection or "ear" of the cartridge. The ear projected up above the breech just high enough to be easily secured for extraction with a finger nail or knife blade, but not high enough to interfere with the line of sight. A glance at the breech would also tell whether or not a cartridge was in the chamber.

The base and projection of the cartridge were stamped out of one piece of sheet brass and soldered to the tube. The projection folded down on itself and was soldered to the rim of the tube, thus forming the complete case. A small hole in the head admitted the flame, from the disk primer to the powder charge.

These rarities were packed in a box with this simple wording:

> The Balls and Brass Tubes should be kept well greased to insure accurate shooting, and to prevent the tubes from sticking in the Gun.

Apparently these cartridges were loaded with two different styles of bullets. At least two well authenticated types were encountered in this research. Both types are illustrated.

CALIBER 45

(45 Whitworth Tube)
Total length $3\tfrac{9}{16}$ in.
Full length heavy paper case .535 dia.
530 gr. cylindrical, round nose, lead bullet
70 grs. black powder

Made in England under English Patent No. 1,959 of 1859. The Whit-

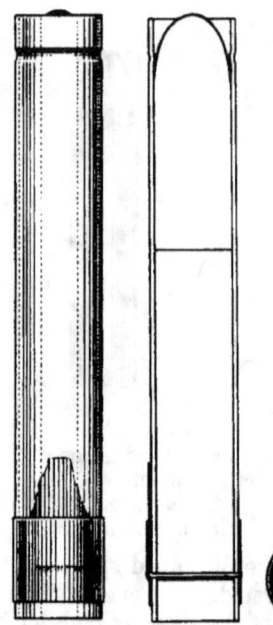

worth tube cartridge was used in the Whitworth rifle. These rifles and cartridges, purchased in England, were smuggled into the South on blockade runners during the Civil War. Used primarily by Confederate snipers, they are said to have given a good account of themselves. A very few of them were equipped with a spiral hexagonal type bullet rather than the cylindrical. These were cast then swaged to the shape of the bore. Their box label reads as follows:

10
WHITWORTH
Patent Cartridges
Cylindrical Projectiles, 530 Grain
Charges 70 grains

The same caliber with 85 grs. of powder was also made . . . length $3\tfrac{29}{32}$ in.

CALIBER 46

(46 Maynard, first type)
Total length $1\tfrac{11}{16}$ in.
$1\tfrac{1}{8}$ in. brass case with cover for base

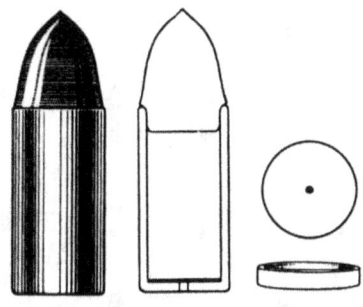

Dia. at head .497
Dia. at mouth .495
Conical lead bullet of .462 diameter

Patented June 17, 1856, (No. 15,141), the cartridge here illustrated is the first Maynard experimental. The cap shown with it was used only to protect the base, in which the pin hole opening into the powder charge was located. A small disk of waxed or gummed paper was placed in the case between the powder and the perforated base. This served as a protection for the powder against air or moisture.

CALIBER 50

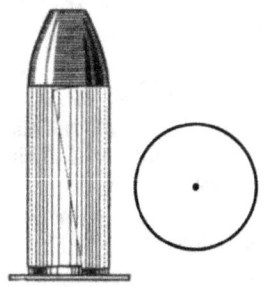

(50 Maynard Brass & Paper)
Total length 1⅝ in.
1³⁄₁₆ in. thin brass, paper wrapped case, .550 dia.
Conical, flat nose, lead bullet

This brass and paper Maynard was used in the early Maynard sporting rifles. It is generally believed that the paper and brass type of case was used previous to the solid brass case.

Be that as it may, this item shows up much less frequent than its near relative, the solid brass case.

CALIBER 50

(50 Smith, heavy paper)
Total length 2¹⁄₁₆ in.
1⁷⁄₁₆ in. heavy paper case
Conical lead bullet

There appears to be a question concerning this and the specimen which follows. It is the opinion of some foremost authorities that these two cartridges are privately made rather than factory issues. Nevertheless, they are encountered by collectors and are pictured along with the others, with only this brief information.

CALIBER 50

(50 Smith's Carbine)
Total length 1²⁷⁄₃₂ in.
1¹³⁄₃₂ in. red paper case with

wording "Smith's Carbine .50"
Conical lead bullet

Like the previous cartridge this one is believed to be of private origin and not regular issue.

CALIBER 50

(50 Smith, rubber case)
Total length 2 1/32 in.
1 1/2 in. black rubber case, .635 dia.
Conical lead bullet for the Smith B. L. percussion rifled carbine

Gilbert Smith of Buttermilk Falls, N. Y., on June 30, 1857, secured a patent (No. 17,702) for a cartridge case . . . "or at least the cylindrical portion thereof, of some impermeable and elastic substance, such as India rubber or gutta-percha . . . so that it may be expanded laterally by the force of the explosion of the charges, and will contract itself after the explosion by its own inherent property." This black rubber cartridge is very seldom encountered nowadays and is a choice collector's item.

CALIBER 50

(50 Smith, brass and paper)
Total length 1 13/16 in.
1 3/8 in. thin brass and paper wrapped case, .630 dia.
Conical lead bullet

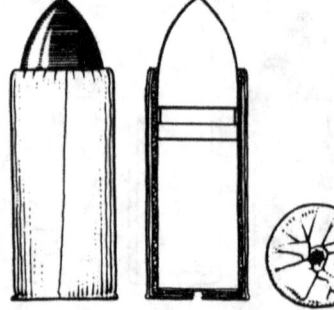

Used in the .50 Smith B. L. Percussion Rifled Carbine

On Dec. 15, 1863, Thomas J. Rodman and Silas Crispin secured a patent (No. 40,988) for a "metallic cartridge case formed of thin wrapped sheet metal . . . combined with an internal or external strengthening disk or cup. Whether this disk is made of paper, metal or elastic material." This patent was assigned to Thomas Poultney at Baltimore, Md. Hence from the label it will be seen that these Smith cartridges were produced under this patent.

10
Poultney's Patent Metallic
CARTRIDGES
Patented Dec. 15th, 1863   12 caps
for
SMITH'S BREECH LOADING
CARBINE
No. 1   50-100 Caliber
Address Poultney & Trimble,
Baltimore, Md.

During the Civil War the government purchased 30,062 Smith's carbines and 13,861,500 cartridges.

CALIBER 50

(50 Maynard with wire extractor)
Total length 2 1/16 in.
1 7/16 in. straight brass case with wire extractor
Conical, flat nose, lead bullet

One of the rare experimentals, this cartridge with its wire extractor was

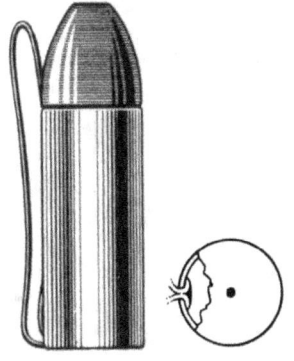

patented Sept. 29, 1863, by Edward Maynard. Patent No. 40,112 provided for a metallic or otherwise durable cartridge with a suitable retracting arm, chain, thong, cord or wire attached. These were used to extract the case when fired. Using this method of extraction, Maynard felt that the necessity of a thick base or build up base was eliminated.

CALIBER 50

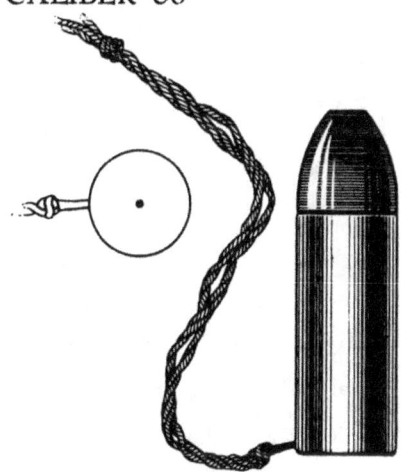

(50 Maynard with Cord Extractor)
Total length 2¼₆ in.
1⅞₆ in. brass, straight case with cord extractor
Conical, flat nose, lead bullet

Another of the rare Maynard experimentals, this one was equipped with a cord extractor. The cord is attached to a small wire loop soldered to the case. On some of these cartridges the cord was passed through a small hole in the case, knotted and then sealed over with wax gum or cement which was not affected by the combustion of the powder.

CALIBER 50

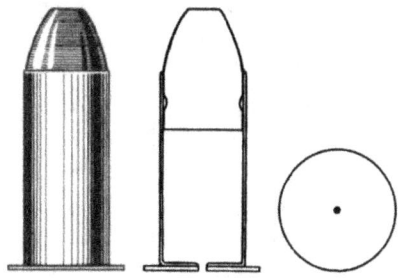

(50 Maynard 1865 issue)
Total length 1¹¹⁄₁₆ in.
1³⁄₁₆ in. brass case .545 dia.
Conical, flat top, lead bullet
Used in the Maynard breech-loading carbine

Best known of the Maynard early brass percussion cartridges, this .50 caliber was used both in the sporting rifles and military arms produced by the Massachusetts Arms Co., Chicopee Falls, Mass., under Maynard's patent. These rifles were the first to use the expanding metallic cartridge case as a gas check.

During the Civil War 20,002 of these rifles were purchased by the government. For use with them 2,157,000 cartridges were purchased.

Saterlee in his *Catalog of Firearms* mentions an interesting item in connection with these arms. It seems that Jefferson Davis secured some 400 of these rifles, probably through a previous War Department purchase, and they were used by the Confederates at Ball's Bluff. Since the factory had burned down, the manufacturers did not get a government contract until 1864.

These rifles will be found both with and without the Maynard tape priming magazine, for which Maynard held the patent.

Occasionally these cartridges are found with the bullet seated quite well down in the case.

CALIBER 50

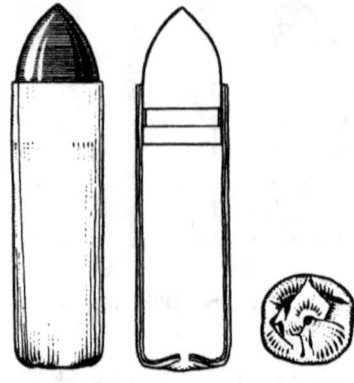

(50 Gallager Brass and Paper)
Total length 2 3/16 in.
1 11/16 in. brass and paper wrapped case
Conical lead bullet

These cartridges were also produced under the Rodman & Crispin patent of Dec. 15, 1863, (No. 40,988). They were used in the Gallager percussion carbine, patented by Mahlon J. Gallager in 1860. Some 8,294,023 of these cartridges were purchased by the U. S. during the Civil War for use in 22,728 carbines.

CALIBER 50

(50 Gallager, All Metal Case)
Total length 2 in.
1 11/16 in. drawn brass, straight case, 550 dia. . . . doughnut type base
Round nose, lead bullet
Used in the Gallager B. L. per-

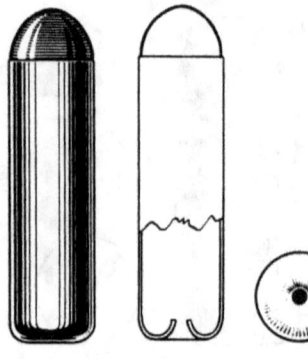

cussion carbine

Comparatively few of these cartridges were made and it is only occasionally that they are found in collections today. In addition to the brass case, they are infrequently found in a tinned metallic case.

CALIBER 50

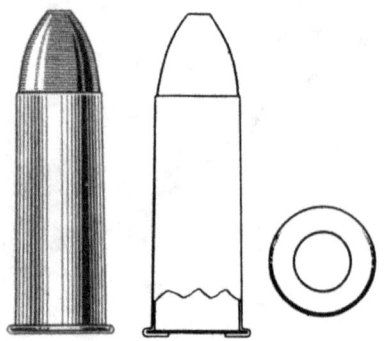

(50 Frankford Arsenal Experimental)
Total length 1 15/16 in.
1 7/16 in. copper, tapering case
Conical lead bullet

One of the early breech-loading, external primed, experimental cartridges. The cross-section shows the nitrated paper base. Such cartridges were, of course, fired by the fire from a percussion cap . . . or tape, if they were used in arms with the latter method of priming.

CALIBER 53

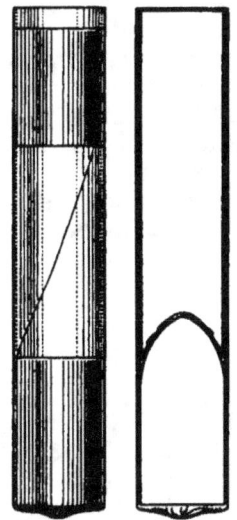

(53 Greene)
Total length 3 in.
Full length paper case
575 gr. cylindrical, round nose, lead bullet
88 grs. black powder

Used in the Greene 1857 Underhammer Bolt Action Rifle and the Greene B. L. Percussion Carbine

The Greene, underhammer bolt action breech-loading rifle used this paper cartridge, having the bullet in the rear, to serve as a gas seal. A bullet was seated in the chamber ahead of the cartridge here illustrated. After acting as a gas seal when the gun was fired, the rear bullet was then pushed forward by the bolt, to be discharged by the powder charge of the next cartridge. The bore of this unusual rifle was smooth but slightly oval in cross section, a feature which its inventor, Lt. Col. J. Darrell Greene adopted from an earlier oval bore gun developed by Charles Lancaster of London, England. The narrow dimension of the bore was .53, the longer dimension .546. The cartridge was patented by Greene on Sept. 8, 1857, (No. 18,143). War Department records indicate that 900 of the rifles were purchased in March of 1863. A similar cartridge was patented in 1860 in England (No. 1,574).

CALIBER 54

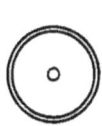

(54 Hunt or Jennings)
Total length 1 in.
Cylindrical, round nose, hollow lead bullet with base plug of cork
Stamped on the side, "Patented 1848"
Diameter .548

Walter Hunt, on August 10, 1848, secured a patent (No. 5,701) for ... "a ball for firearms, with a cavity to contain the charge of powder for propelling said ball, in which cavity the powder is secured by means of a cap enclosing the back end." A reissue, No. 164, dated Feb. 26, 1850, further describes this unique cartridge ... "Making metallic balls for firearms with the rear part thereof cylindrical and a cavity in the said cylindrical part of sufficient capacity to receive the entire charge of gunpowder, substantially as herein described, when the said charge is retained in the ball by a cap or the equivalent thereof, having a central hole through which the charge can be inflamed."

Twenty of these cartridges were carried in a tubular magazine below the barrel of a rifle patented by Lewis Jennings in 1849. They were fired by a Maynard tape-primer held in the priming magazine on top of the frame.

Tyler Henry adapted the ideas of this

gun and those found in the Horace Smith patent of 1851 (Volcanic) into the famous Henry Magazine Rifle of 1860.

Due to the short life of the Jennings arms, few of the Hunt cartridges are to be found today.

A similar cartridge was patented in England in 1847 ... (No. 11, 994). It was described as "Powder in projectile closed in by perforated metal plate."

CALIBER 54

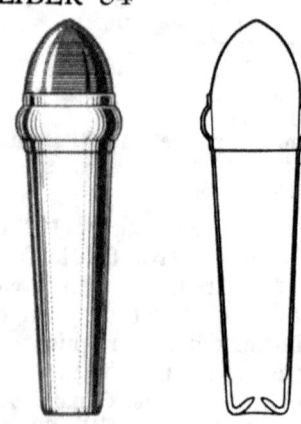

(54 Burnside)
Total length 2⅜ in.
1¹⁵⁄₁₆ in. brass tapering case with large grease ridge at mouth
400 gr. conical lead bullet
45 grs. black powder

The Burnside cartridge is one of the earliest of American brass cases. It was used extensively in the Burnside B. L. Carbine during the Civil War and later, up until the time of the Custer Massacre.

On March 25, 1856, Ambrose E. Burnside secured a patent (No. 14,-491) on a metallic tapering case with a perforation in the base by means of which the powder could be ignited from the fire of a percussion cap. George P. Foster, on April 10, 1860, secured a patent (No. 27,791) for a grease chamber around the bullet of the Burnside cartridge. It is this later cartridge which is quite common today, and of which the Ordnance Department purchased the large quantity of 21,819,200 during the Civil War. It is understood that this cartridge was also made in a 2⅛ in. case with a .50 caliber bullet, but it is a very rare item indeed. An original label from a box of these unusual cartridges reads:

10 CARTRIDGES
with 12 caps
for the
B U R N S I D E
BREECH LOADING RIFLE
Patented March 25, 1856
Caliber 54/100
Made by
Burnside Rifle Co.
Providence, R. I.

CALIBER 56

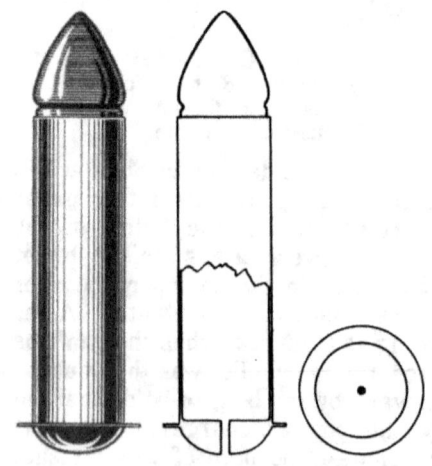

(56 Billinghurst Requa Mitrailleuse)
(Pronounced Me-try-youz)
Total length 2⅝ in.
2 in. brass, straight case .582 dia.
Conical lead bullet

Used in the Billinghurst Requa Volley Gun

Even though this cartridge was used in a wheel mounted gun having twenty-four barrels in a single row, it is included in this digest because of its unusual interest in collections. The cartridges were loaded in the chambers simultaneously, 24 cartridges on a long, hinged, metal feed strip. The gun was one of the first of the quick firing multi-shot arms. A train of priming powder reaching the full length of the breech block was set off by a percussion cap. This in turn ignited the powder, through the small hole in the base of each cartridge, setting off the whole row in short order.

CALIBER 58

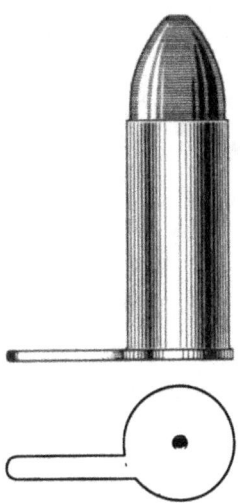

(58 Maynard "Flop Ear")
Total length $2\frac{1}{16}$ in.
$1\frac{7}{16}$ in. brass, straight case
Conical lead bullet

A further type of one of the very rare early Maynard experimentals. This type was included under Patent No. 40,112 of Sept. 29, 1863. The flop ear is for extracting the fired case from the chamber.

CALIBER 69

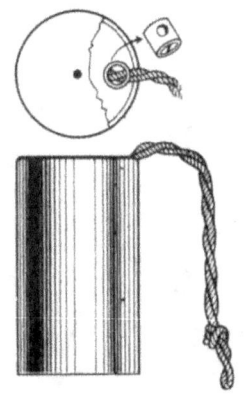

(69 Maynard with Cord Extractor)
$1\frac{5}{16}$ in. brass, straight case

It will be noted in this Maynard experimental that the cord is attached to the interior of the case by means of a set screw in a small bushing.

CALIBER 69

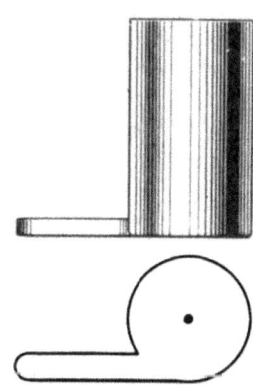

(69 Maynard "Flop Ear")
$1\frac{5}{16}$ in. brass straight case

This .69 Flop Ear differs from the .58 Flop Ear in that the flop ear on the .58 cartridge is set in a short way from the edge . . . while on this size the flop ear is tangent with the perimeter of the base.

CALIBER 69

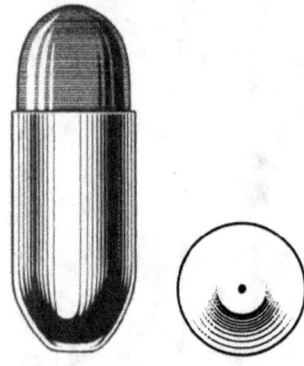

(69 Maynard Capsule Shape)
Total length 2 in.
1⅜ in. brass, tapered base case
Cylindrical, round nose, lead bullet

A Maynard experimental about which little is known. Since only a few were made for experimental purposes, they are quite rare indeed today and are to be found only in a few very extensive cartridge collections.

self-contained

PAPER

METALLIC

SELF CONTAINED

Had Napoleon envisioned the possibilities of the bolt action rifle offered to him by M. Pauli, a Swiss, it is possible the course of history might have been altered ... not alone by the rifle, but by the self-contained cartridge made for it. In a way similar to a modern shotgun shell, it had the ball and powder contained in a combustible case. This unit was seated in a brass base in which was placed a detonating pellet in a primer pocket. This head could be reloaded again and again. But Pauli's invention was too far ahead of its time to be appreciated.

Still the experiments went on to develop a cartridge which would be a complete unit. Needle fire cartridges in which the powder, bullet and fulminate were all contained in one unit, were introduced. They were the first complete cartridge to be in general use. Some had the fulminate ahead of the powder and some had it in the base, as the French Chassepot.

An auxiliary chamber provided with a nipple for a percussion cap was a development of this time. It was used both in long arms and hand arms.

Between 1847 and 1850 impetus was given cartridges of this type by M. Houillier. A series of his patents covered the use of metal for complete cartridge cases, irrespective of the type of primer used.

Truly the separated units of the cartridge were giving way to the new ... complete unit.

CALIBER 7 mm.

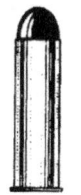 

(7 mm. Cane Gun)
Total length 1 3/32 in.
2 9/32 in. paper case with brass head . . . .276 dia.
Round nose, lead bullet

Picked up in Mexico, this cartridge was a type used in the early cane guns. The cross-section illustrates how the brass head containing primer was set into the paper case.

CALIBER 31

(31 Volcanic)
Total length 11/32 in.
Cylindrical lead bullet, .322 dia. . . . with either cork base or brass and cork base
Used in Volcanic hand guns

Volcanic arms using a self-contained cartridge mark a definite period in arms and ammunition development. Not only were they the daddy of the Henry rifle, the grand-daddy of the famous Winchester lever-action rifle . . . but it was they who signaled the end of the percussion era.

An interesting angle on the Volcanic cartridges came to light during this research. The Volcanic arms were patented Feb. 14, 1854, and were awarded the gold medal at the Maryland Institute Exposition in 1855. Yet . . . and here is the interesting angle. The Volcanic cartridge was not patented until January 22, 1856, (Patent No. 14,147). On a box of the .41 cal. size, the patent date of Aug. 8, 1854, is given. This patent, No. 11,496, is for an entirely different type of cartridge . . . "A copper shell with metal disk near base to serve as anvil, and metal disk back of bullet to hold tallow next to the ball." This patent was held by Smith & Wesson. While the No. 14,147 was assigned to the Volcanic Repeating Arms Company . . . who had purchased the manufacturing rights for Volcanic repeating arms from Smith & Wesson.

So, while Jan. 22, 1856, was the actual patent date for the Volcanic cartridges, they were really being produced under another patent for a different type of cartridge.

In 1854 an English patent, No. 424, was granted for a . . . "Projectile containing charge. Disk in powder carries fulminate." The patent drawing on this indicates a close similarity to our Volcanic. (See plate 109, *Hand Cannon to Automatic*)

CALIBER 8 mm.

(8 mm. for Gravity-Fed Pistol)
Total length 11/32 in.
Cylindrical, conical pointed, lead bullet

This unusual cartridge is similar to the 9 mm. Gaupillat. It is smaller, with a shorter conical point. The copper primer sets up in the base of the bullet rather than being flush with it. While the specimen examined was not taken apart, it is doubtful if this tiny bullet contained more than fulminate for a propellant.

CALIBER 34

(110 bore Eley's Needle Gun)
Total length 1⁵⁄₁₆ in.
²⁹⁄₃₂ in. paper case with dark blue base
Conical lead bullet

The title "needle gun" came about because of the resemblance of the firing pin to a needle. In discharging the cartridge, the needle penetrated the head of the cartridge to strike the fulminate, which in this case was contained in a percussion cap in the head—rather than near the base of the bullet. Only a few of the needle-fire type of guns were ever manufactured in America.

CALIBER 9 mm.

(9 mm. Gaupillat)
Total length ½ in.
Cylindrical, conical point, lead bullet

Used in the gravity-fed pistol, this unusual cartridge was patented in France (No. 10,698) in 1854, by a Mr. Gaupillat. It is very similar to the American Volcanic in that both primer and powder are contained in the hollow base of the bullet.
The unusual pistol using this cartridge was patented by a Mr. Collette in the early half of the 19th century. (See plate 110, *Hand Cannon to Automatic*)

CALIBER 38

(90 Bore Eley's Needle Gun)
Total length 1⅜ in.
¹⁵⁄₁₆ in. paper case with orange base
Conical lead bullet

Another of the needle gun cartridges manufactured by the English firm of Eley Bros. of London.

CALIBER 38

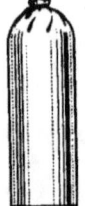 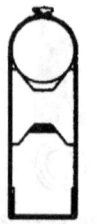

(.38 Needle Fire)
Total length 1¼ in.
Rimless paper case tied at the top . . . .407 dia.
Spherical lead ball
Used in a light rifle

These interesting cartridges were first written up in "The Gun Report" of April, 1942. They were brought back from Europe, where they had been discovered in their original box by an American collector. The label on the box reads:

25 STUCK PATRONEN
KUGEL
Zundnadel Teschin
Mit Hebel und Excenter
Caliber 0.38

The cross-section of this rare find gives a good idea of the construction of a needle fire cartridge.

CALIBER 38

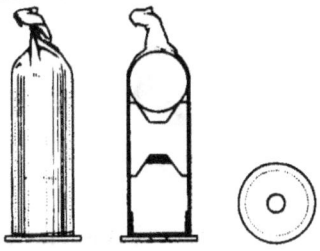

(38 Needle Fire)
Total length 1⅜ in.
Rimmed paper case, .378 dia.
... twisted at the top
Spherical lead ball
Used in a light rifle

When found, these cartridges were in a box wrapped only in coarse brown paper.

Between the powder charge and the perforated base is a piece of thin paper. Ahead of the black powder charge is a cardboard wad. In the concave under side of this wad is located the priming pellet ... which, when struck by the needle, detonated the powder charge.

CALIBER 38

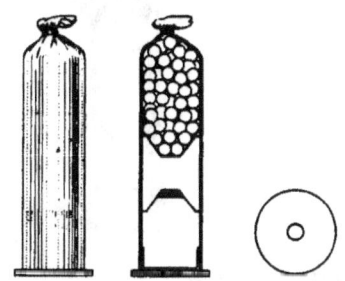

(38 Needle Fire Shot)
Total length 1½ in.
Rimmed paper case, .373 dia.
... tied at the top
Lead shot pellets
Used in a light shot gun

The third type of the foreign needle-fire cartridges brought back from Europe in 1937.
They are described by the label on the original box as ...

25 Stuck Patronen
fur
Zundnadel Schroot Techin
mit Pappverschluss
Caliber 0.38 ... 9.95 mm.

CALIBER 41

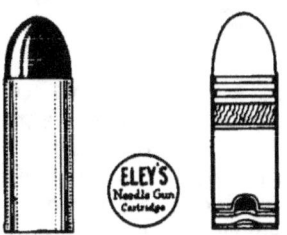

(75 Bore Eley's Needle Gun)
Total length 1¼ in.
²⁹⁄₃₂ in. rimless paper case
Conical, round nose, lead bullet

The box label reads as follows:

50
ELEY'S
NEEDLE - RIFLE
CARTRIDGES
75 Bore

To insure ignition, the needle of the rifle should strike $\frac{3}{16}$ of an inch through the breech, and should be kept sharp and free from rust.

ELEY BROS. MANUFACTURERS, LONDON

The cross-section view illustrates the location of the fulminate in this type of needle-fire cartridge.

CALIBER 41

(41 Volcanic) (made also in a 36 cal.)
Total length 1¹¹⁄₁₆ in.
Cylindrical lead bullet, .415 dia. ... round nose, and with either cork base—or cork and brass base
Used in Volcanic hand guns and rifles

In order to clear up some misinformation which has existed and which has been passed along by recent

writers, a detailed cross-section of this cartridge is shown. A glance at it will show that, contrary to popular conception, the detonator and propellant were not in one mixture ... but were separate and distinct units.

Quoting now from the patent papers, No. 14, 147 ... "Operation: the ball being placed in the arm, a blunt projecting piece is pressed through the cork until it bears on the fulminating powder and disk. A smart blow from a hammer will then ignite the percussion, and by forcing the fire through the openings a, a, a, a, Fig. 6, will explode the powder in the ball." An examination of an actual cartridge verified this explanation.

The label from an original box reads thus:

        200 -------- No. 2
        PISTOL CARTRIDGES
          Manufactured by
     THE VOLCANIC REPEATING
            ARMS CO.
          New Haven, Conn.
        Patented Aug. 8, 1854

CALIBER 41

(41 Williamson Derringer Auxiliary)
Total length 7/8 in.
Steel case with nipple for percussion cap

D. Williamson, on Oct. 2, 1866, secured a patent (No. 58,525) for this metallic case with a nipple. It was developed for use with his derringer ... a gun which could use .41 caliber rimfire ammunition ... or this auxiliary chamber loaded with powder and ball and equipped with a percussion cap.

Rumor has it that "Wild Bill" Hickok, of frontier fame, carried a pair of these unique arms. Irrespective of this, they were one of the few combination cartridge and percussion small arms made. (See plate 129, *Hand Cannon to Automatic*)

CALIBER 41

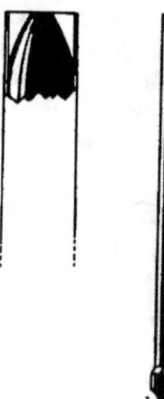

(41 Roper)
Total length of iron case $2^{13}/_{32}$ inches
  Dia. at mouth .466
  Dia. above flange .476
  Dia. of flange .545
Flat nose lead bullet

Roper shotguns and rifles were the inventions of S. H. Roper, and were patented April 10, 1866, (No. 53,881). Three firms manufactured the arms.

Hopkins & Allen Mfg. Co., Norwich, Conn.
Roper Sporting Arms Co., Hartford, Conn.
Roper Repeating Rifle Co., Amherst, Mass.

Roper rifle cartridges are to be found only in a very few collections. They are quite scarce and not a great deal of information is available concerning them. It is generally agreed that the original Roper cartridge had a flat nose bullet seated flush with the mouth of the case as illustrated in the cut-away sketch.

The Roper revolving rifle had a cylindrical receiver which contained a central revolving magazine with four divisions. Each division held one of these metallic cartridges. When the hammer was cocked, after the arm had been fired, it extracted the fired shell from the barrel and returned it to the division formerly holding it . . . and in turn revolved the magazine until another division was in line with the barrel. When the trigger was pulled, it drove the hammer forward together with a combination bar, firing pin and extractor, to push the cartridge into the barrel and fire it.

Cartridges were removed or loaded into the magazine by means of a hinged door on top side of the receiver covering the revolving magazine. The length of the opening of this trap door is approximately 2¼ in. It was therefore necessary to tilt the cartridge to permit its loading into the magazine.

Roper arms were available with both rifle and shotgun barrels, having magazine covers attached. The magazine itself however, had to be large enough to carry either rifle or shotgun cartridge.

CALIBER 43

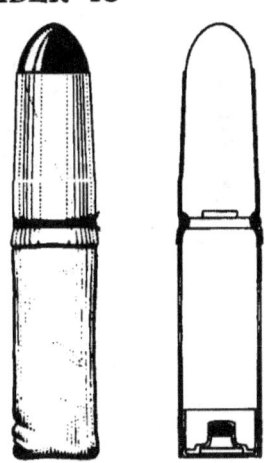

(11 mm. French Chassepot)
Total length 2 9/16 in.
Paper case tied in middle with cord
380 gr. elongated, conical, round nose, lead bullet
85 grs. black powder
Used in the Chassepot B. L. Military Rifle

This is the cartridge used by the French in the Franco-Prussian War of 1870. It consists of six main parts, powder case, powder, pasteboard, bullet case, bullet and primer. The primer is composed of a copper cap, like an ordinary military percussion cap, only a bit smaller. Formed at the bottom, it has two holes directly opposite each other to permit the flame from the fulminate to reach the powder charge. The cap is covered with a thin disk of brass, copper or other metal which is attached to the paper forming the base of the cartridge.

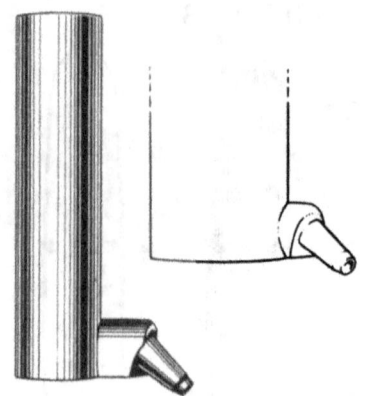

There is a possibility that this rifle cartridge, and the shot shell detail shown with it, may have been used in a gun having interchangeable barrels. It is a good example of an auxiliary chamber for long arms and is illustrated for its interest.

CALIBER 51

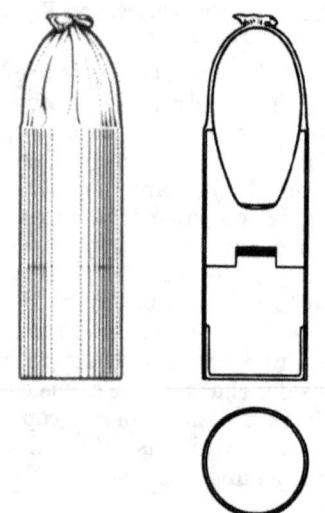

(13 mm. Prussian Needle Gun)
Total length 2 3/16 in.
Paper case tied at top
530 gr. egg shape, lead bullet
70 grs. black powder
Used in the German Needle Gun—an invention of Johann Nicholas Von Dreyse

Von Dreyse invented the needle gun in 1827, but it was not until 1841 that a few were issued to German troops for tests. It was the granddaddy of all modern bolt-action rifles. In 1864 the Prussians used it against Denmark, in 1866 against Austria and against France in 1870.

The cartridges for this interesting arm are quite scarce in this country, due to the fact that at the time of their use no ship cared to carry them with their more or less fragile paper cases.

The cross-section view gives a very good idea of their construction. As will be noted the bullet is seated in a heavy papier mache' cup or sabot. The fulminate is contained in a small indentation on the under side of this sabot, below this is the powder charge. To fire the cartridge, the needle must pierce through the powder charge to strike the fulminate.

The specimen illustrated is the third type (1855) having the egg-shaped bullet . . . or should we say . . . "streamlined."

It will be observed that the bullet is smaller than the paper sabot. The sabot, being soft and malleable, is compressed, upon firing, to fit the grooves, preventing the escape of gases. It also sets the bullet to spinning by following the rifling.

The Prussian needle gun cartridge is one of the really historic cartridges. It was one of the first to be entirely self-contained . . . case, powder, bullet and primer.

CALIBER 58

(58 Auxiliary Chamber)
Total length 2 3/4 in.
Iron case .964 outside dia.
.595 inside dia.

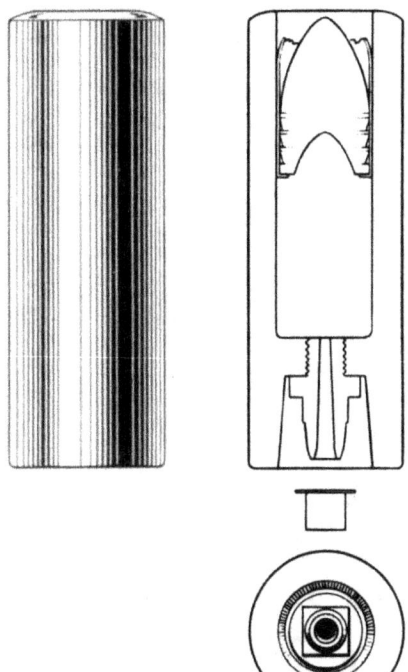

tightly against the rear of the barrel to prevent the escape of gas. After the firing, the cylinder carried the empty chamber pieces around to an opening which allowed them to fall out. They could then be reloaded and used again indefinitely. Rapidity of fire was dependent upon the number of chamber pieces that were loaded before the battle. The gun in the West Point Museum was used by Union Troops at the siege of Petersburg, Virginia during the Civil War. (Gerald C. Stowe, Curator, West Point Museum).

This chamber piece was used in the Union Repeating Machine Gun, caliber .58, made by E. Nugent of New York. This is the earliest machine gun known to have been used by the United States Army. Fifty-one of the guns were purchased in 1862. The main drawback which hindered the development of the machine gun was the lack of a self-contained metal cartridge. This gun solved the problem by utilizing a number of short muzzle-loading chamber pieces, each with a percussion nipple at the rear. A number of these pieces were loaded and capped ahead of time and later fed by means of a hopper into a fluted cylinder which revolved and brought the charges in succession to the barrel, pausing long enough to allow the charge to be fired. A wedge shaped block forced each small chamber piece

patent ignition

- PIN FIRE
- LIP FIRE
- ANNULAR RIM
- CUP PRIMER
- TEAT FIRE
- INSIDE PIN FIRE

PATENT IGNITION

Into this class fall a number of cartridges of a more or less odd design. Oldest and by far the most generally used is the pin fire, a type which is still in considerable use in Europe and South America. It had a pin protruding through the side of the case near the head. The point of the pin on the inside rested in a percussion cap containing fulminate, which was exploded when the pin received a blow from the hammer.

An inside pin fire, the Gallager & Gladdings, was patented in this country for use in a long arm, but few specimens of it are in existence.

One picturesque cartridge much sought after by collectors is the unique Crispin with its fulminate contained in an annular ridge midway up on the case.

The lip fire, teat fire and cup primer types were more of the odd types designed to circumvent the patents held by Smith & Wesson. One of these patents was that which provided for the boring clear through of the cylinder chamber to permit the loading of a cartridge from the breech end.

Various calibers were made in most of these systems, both for long and short arms. Truly an interesting period . . . bringing out man's ingenuity to produce something different.

CALIBER 5 mm.

(5 mm. Pin Fire)
Total length 21/32 in.
7/16 in. copper straight case, .226 dia.
Cylindrical, pointed, lead bullet

An unusual cartridge which never fails to attract anyone even remotely interested in cartridges. It was made in a variety of calibers for both long and short arms.
The label from a typical box is as follows:

          100
    5 mm. Pin Fire
    CARTRIDGES
    for Revolvers
ELEY BROS., Ld. London

CALIBER 25

(25 Allen Lip Fire)
Total length 1 3/16 in.
1/2 in. copper straight case, .257 dia.... with projecting lip for fulminate
Conical lead bullet
Used in the Allen & Wheelock 7-shot cartridge revolver

On Sept. 25, 1860, a patent (No. 30,109) was granted to E. Allen for a metallic cartridge having a side teat for fulminate.

An old label reads:
Cal. 25   No. 50   Allen Lip Cartridge
          Water-Proof
   PISTOL CARTRIDGES
Allen's Pat.   Sept. 25th, 1860

CALIBER 28

(28 Front Loading Cup Primer)
Total length 15/16 in.
Copper case, .266 dia.—with flange at mouth and concave base
Conical lead bullet

Willard C. Ellis and John C. White secured a patent (No. 24,726) on this front-loading cartridge on July 12, 1859. The obvious idea back of it was to circumvent the patent held first by Rollin White and later purchased by Smith & Wesson. This patent provided for the b o r i n g through of a cylinder in order to load a cartridge from the breech end. Three years later (Aug. 25, 1863) they assigned a similar patent to Ebenezer H. Plant, Henry Reynolds, Amzi P. Plant and Alfred Hotchkiss. (See plate 117 *Hand Cannon to Automatic.*)

CALIBER 31

(31 Crispin)
Total length 1 1/8 in.
23/32 in. copper case, .326 dia. —with annular rim, .391 dia. for fulminate
Conical lead bullet
Used in the Crispin revolver, patented by Silas Crispin, Oct. 3, 1865

Silas Crispin, on Aug. 8, 1865, was granted a patent (No. 49,237) for this unusual cartridge. The patent papers describe it as follows: "The cartridge constructed as described, that is to say, with the fulminate placed within a projecting annular recess or rim, which is formed at a point between the ends of the cartridge case." (See plate 127 *Hand Cannon to Automatic*)

CALIBER 32

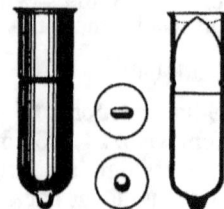

(32 Teat Fire)
Total length 1¼ in.
Copper case, .343 dia.—with flange at mouth and teat containing fulminate at base
Conical lead bullet
Used in the National Front-Loading revolver

Another of the ingenious cartridges developed to get around the Smith & Wesson patent. This one was patented Jan. 5, 1864 (No. 40,183), by D. Williamson. It was advertised in the May 21, 1864, issue of the Army and Navy Journal.

As will be observed from the sketches they were supplied in two styles, one with a round teat and one having a flat teat. A label from a box of the round teats reads as follows . . .

ROUND TEAT No. 2
Fifty No. 32
Metallic
Central Fire Water-proof Cartridges
for
IMPROVED NATIONAL REVOLVER
Manufactured by the
National Arms Co., Brooklyn, N. Y.

CALIBER 42

(42 Front-Loading Cup Primer)

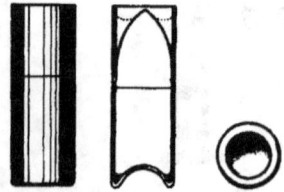

Total length 1³²⁄₃₂ in.
Copper case, .410 dia.
Conical lead bullet
Used in the Plant Army revolver

The case of these cartridges reached the full length of the chambers in the cylinder. The hammer tip struck the inside concave portion of the base, in which the fulminate was contained, to detonate the powder charge.

CALIBER 44

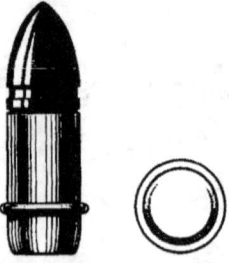

(44 Crispin)
Total length 1⁹⁄₁₆ in.
⅞ in. copper, or brass case
Cylindrical, pointed, lead bullet

One of the very scarce items in the Crispin family of cartridges — and one which is found in only a very few collections.

CALIBER 44

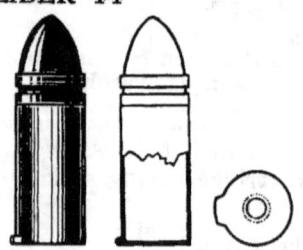

(44 No. 58 Allen Lip Fire)

Total length 1⅞₁₆ in.
⅞ in. copper, straight case, .450 dia.
Conical lead bullet

This is the largest size Allen Lip Fire cartridge ... the other two being .25 and .32 caliber sizes. It was used in the large Allen & Wheelock six-shot revolver. The cross-section view illustrates the position of the fulminate in the "lip." (See plate 114 *Hand Cannon to Automatic*)

CALIBER 45

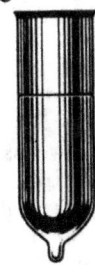

(45 National Teat)
Total length 1½ in.
Copper case, .474 dia.—with round teat
Conical lead bullet
Used in the National Front-Loading revolver

This scarce cartridge was more or less unknown until quite recently. Even then there was some question as to the actual caliber. The following wording taken from one of the original boxes should clear up any doubt.

25 No. 45
ARMY AND NAVY
NATIONAL REVOLVER
CARTRIDGES
Manufactured by
THE NATIONAL ARMS CO.
Brooklyn, N. Y.

CALIBER 12 mm.

(12 mm. Pin Fire ... short)
Total length 1 in.
⁹⁄₁₆ in. copper case, .462 dia.
—may also be found with

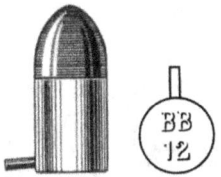

brass case.
Conical lead bullet
Manufactured by Braun & Bloem, Germany

Pin fire cartridges, like other types, were made in several caliber sizes, case lengths, etc. The one illustrated is a short produced by one of the better known German firms. Several European manufacturers produced these types of cartridges.

CALIBER 12 mm.

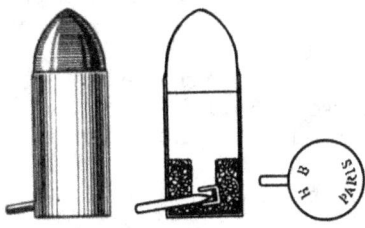

(12 mm. Pin Fire)
Total length 1¼ in.
⅞ in. copper case, .462 dia.
Conical lead bullet
Manufactured by Houllier & Blanchard, Paris

Illustrated with this cartridge is a cross-section view. It will describe better than words the operation of a pin fire. A blow from the hammer drives the pin against the fulminate in the percussion cap, thus detonating it and in turn igniting the powder charge.

CALIBER 50

(50 Crispin)
Total length 1³¹⁄₃₂ in.
1½ in. copper case, .558 near rim ... annular rim one-third up from base, .675 dia.
Conical, flat top, lead bullet

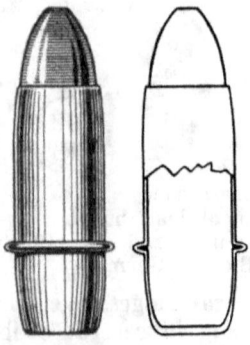

Even though it is the most common of the Crispins, it is one of the most sought after by cartridge collectors. It is unique and its appearance never fails to draw the attention of those interested in cartridges. Patented Aug. 8, 1865, (No. 49,237) ... it was used in a special model of the Smith B. L. Carbine. This cartridge was also made with a shorter case. Crispin secured an English patent, No. 3,258, in 1865 for this same cartridge.

CALIBER 58

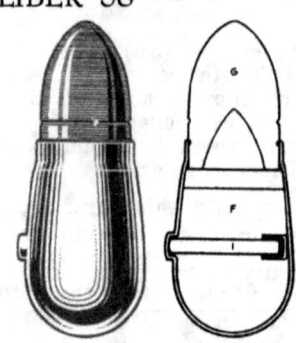

(58 Gallager & Gladding)
Total length $1^{29}/_{32}$ in.
$1^{3}/_{16}$ in. brass case, .680 at widest part . . . .652 at mouth
Cylindrical, pointed, lead bullet

Rarest of the rare is this inside pin fire patented July 12, 1859, (No. 24,730) by Gallager & Gladding. The patent provided for "(Breech Loader) Case of paper or wood, or metallic shell, or wood or paper banded with metal. Pin fire."

The August 31, 1861, issue of *The Scientific American* has this to say of this cartridge . . . "The cartridge case, *f*, is formed of thin brass, and is attached to the Minie' bullet, *g*, by pressing the rear of the latter into the mouth of the cartridge case; when the joint is made water-proof by dipping the cartridge in melted tallow. A cork wad is interposed between the powder and bullet to clean out the gun at the discharge. As the cartridge is entirely closed it is necessary to introduce percussion powder (fulminate) to the inside in order to fire the powder. This is effected by placing a common percussion cap on one end of a small wire, *i*, which is then fixed across the cartridge at its greatest diameter. A small blister is formed on the cartridge at the end of the wire, and a corresponding cavity is made at the breech of the gun to guide the cartridge at its introduction so as to bring the end of the wire directly under the cock."

As mentioned, Gallager & Gladding secured the patent on this cartridge in 1859. Yet the *Scientific American* article, part of which is quoted above, describes the Schubarth cartridge. It would seem that Schubarth was claiming the Gallager & Gladding patent since there is apparently no record of such a patent ever being granted to Schubarth. It may be that if the last paragraph of this *Scientific American* article is correct, that Schubarth was late in seeking a patent. "One patent for this admirable invention (the gun) was granted July 23, 1861, and applications for other and for the cartridge have been made through the Scientific American Patent Agency."

A specimen of the Schubarth breechloading rifle is in the Smithsonian.

Could it be that Gallager & Gladdings patented the cartridge for use in Schubarth's rifle? A Casper D. Schubarth of Providence, R. I., was a Civil War contractor of percussion muskets.

CALIBER 15 mm.

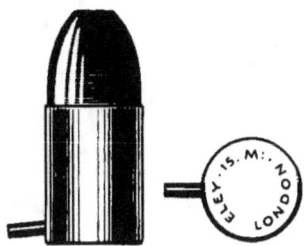

(15 mm. Pin Fire)
Total length 1⅜ in.
1³⁄₁₆ in. brass case, .616 dia.
 (also made in copper case)
Cylindrical, flat nose, lead bullet

This is the largest pin fire revolver and carbine cartridge made. It is described briefly by the label on the box as:

50 ELEY'S 15 mm.
15 mm. Pin Fire
CARTRIDGES
For Carbines & Revolvers
Mfg. by
Eley Bros. - Limited London

rim fire

COPPER

BRASS

RIM FIRE

When Flobert of France invented the bullet breech cap, he started a type that has not only survived, but is immensely popular even today. Differing only from the earlier issues by the addition of powder, a longer case and a wide range of calibers, the present day rim fires have changed little from their first ancestors in appearance.

In this country Smith & Wesson pioneered by perfecting a revolver to use rim fire cartridges . . . their hinged frame .22 caliber revolver.

While the Flobert cartridge was designed only for indoor target shooting, the early S & W was clearly intended for a more serious purpose, that of protection and utility.

B. Tyler Henry was the first to see the possibilities of this new cartridge for larger calibers. He produced a .44 caliber size which was used in the Henry magazine rifle, forerunner of the famous Winchester line.

Rim fire cartridges were used in both military and sporting rifles, revolvers and carbines.

The first conversions of the percussion muskets of the Civil War to metallic cartridges utilized this then new cartridge. The .58 caliber, one of the largest rim fire cartridges made, is quite a contrast to the earlier .22 short so far as size is concerned.

It is not an exaggerated statement to say that in all probability more .22 caliber rim fire cartridges have been made than any other single cartridge. It is so common that it is known by every schoolboy.

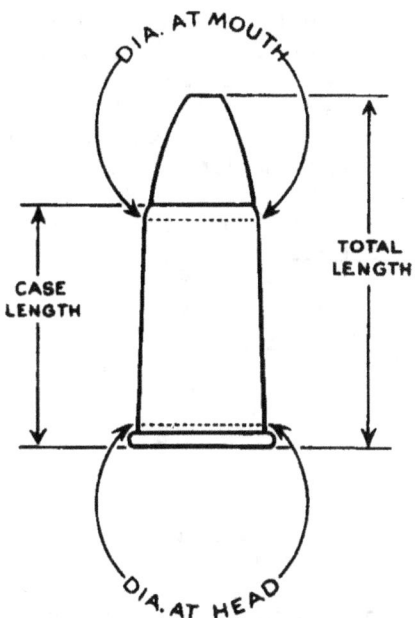

CARTRIDGE MEASUREMENTS

The diagram above explains how the cartridges were measured for this digest. On cartridges where the case was straight and not crimped into the bullet, the "diameter at mouth" was taken further up or right at the mouth. On rimless types, the case was miked just above the groove.

CALIBER 2 mm.

(2 mm. Rim Fire)
Total length ⅛ in.
³⁄₃₂ in. copper case
Spherical lead ball

Smallest rim fire cartridge known. It is used in the miniature European made watch charm revolvers which were advertised some years ago by Stoegers of New York.

CALIBER 4 mm.

(4 mm. Genschow)
Total length ⁵⁄₁₀ in.
¼ in. copper case
Spherical lead ball

This is the type of cartridge used in the European saloon rifles (Zimmerstutzen). These rifles look much the same as any European target rifle except that the cartridge is loaded through a small hole a few inches back of the muzzle. The rest of the barrel carries the firing pin plunger. The acorn on the head is the mark of Gustav Genschow, Germany.

CALIBER 14

(14 Jones Express)
Total length ²⁵⁄₃₂ in.
¹⁹⁄₃₂ in. brass bottleneck case
    Dia. at head .225
    Dia. at mouth .167
Cylindrical, flat nose, lead bullet

This is what is commonly referred to among shooters and collectors as a "Wildcat." That is to say it is not a commercial product, but is one which is made by a private individual for some certain gun which has been bored and rifled in an odd size caliber. This particular cartridge was made up from a UMC .22 long case, necked down to .14 caliber. It was used in a rebored 3½ in. barrel Stevens "Tip-Up" model pistol.

CALIBER 22

(22 BB Cap)
Total length 1³⁄₃₂ in.
¼ in. copper case, .226 dia.
20 gr. conical, round nose, lead bullet

Commonly known as the BB cap or "bullet breech cap," this cartridge had its origin back around 1845 when Flobert of France invented such a cartridge to use in the early saloon target rifles. It was originally loaded with a round .22 caliber ball, no powder charge, the ball being driven by the fulminate in the rim.
It is now loaded with a very light charge of smokeless powder.

CALIBER 22

(22 C. B. Cap)
Total length 1⁷⁄₃₂ in.
¼ in. copper case, .225 dia.
29 gr. conical lead bullet

"Conical Bullet Cap" is the correct name for this cartridge, which is used, for the most part, for indoor gallery practice shooting. It was originally loaded with a black powder charge . . . then a semi-smokeless charge . . . and now with smokeless powder. It uses a slightly heavier charge than the BB cap.

CALIBER 22

(22 short)
Total length $1\frac{1}{16}$ in.
$1\frac{3}{32}$ in. copper case, .225 dia.
29 gr. conical, lead bullet
4 grs. black powder

Here is the best known of all American cartridges. It is the oldest in point of service of cartridges of American origin. Originally developed by Daniel Wesson for the Smith & Wesson First Model revolver, which came out late in 1857, its outside appearance has undergone but little change in all the years since.

A variety of bullets are available today—the plain solid lead, outside greased . . . plain lead, dry wax coated . . . cadmium plated . . . copper plated, (these latter two come either dry, wax coated or grease lubricated) . . . and the hollow point. Case may be either copper, brass or nickeled.

More of these cartridges have been produced than any other single known cartridge. (See plate 111, *Hand Cannon to Automatic*)

CALIBER 22

(22 BB Cap . . . foreign)
Total length $\frac{3}{8}$ in.
$\frac{1}{4}$ in. copper case
Spherical lead ball
Used in indoor target pistols and rifles

This specimen was brought back from Argentina. The bird (Pajarito) is the trademark of Spreafico, a cartridge manufacturer of Buenos Aires.

This same firm also uses the initial SAS as a headstamp on some of their cartridges.

It is interesting to note that the firm of Braun & Bloem, Germany, also uses a similar bird on their cartridges.

CALIBER 22

(22 Unknown)
Total length $1\frac{1}{32}$ in.
$2\frac{1}{32}$ in. brass case, .225 dia.
Cylindrical lead bullet

The head carries the familiar U of Rem-UMC . . . but beyond that, nothing is known of this unusual cartridge. It is included in this digest merely for its unique appearance. Perhaps it is one of the many wildcats . . . but what type of gun was chambered for it?

CALIBER 22

(22 Extra Long)
Total length $1\frac{5}{32}$ in.
$\frac{3}{4}$ in. copper case, .225 dia.
40 gr. cylindrical lead bullet
6 grs. powder
Manufactured by Rem-UMC especially for Winchester, Maynard, Stevens and other rifles of past years

One of the rim fire cartridges which may be only a collector's item in a few years. It is being pushed aside by the popular .22 long rifle cartridge.

CALIBER 6 mm.

(6 mm. Gaulois Salon)
Total length 15/32 in.
9/32 in. two piece, brass and copper case
  Dia. at head .232
  Dia. at mouth .232
Pointed, lead bullet

One of the European target cartridges used for indoor shooting. This type of cartridge used no powder, but employed an extra amount of fulminate which served as a propellant.

The brass base of this cartridge fits over the thin copper case which extends on up to hold the bullet.

CALIBER 25

(25 Bacon & Bliss)
Total length 25/32 in.
15/32 in. copper case, .245 dia.
43 gr. cylindrical lead bullet
5 grs. black powder

This cartridge was developed for the F. D. Bliss Cartridge Revolver, a gun which was an infringement of the Smith & Wesson patent. It was discontinued around 1863.

A label from the box reads:
  100 .25 cal. SHORT
  RIM FIRE CARTRIDGES
 Swaged   Bullets
  Manufactured by
UNION METALLIC CARTRIDGE CO.
  Bridgeport, Conn.
   U. S. A.

CALIBER 25

(25 Long for Stevens)
Total length 1 3/8 in.
1 3/32 in. copper case, .277 dia.
67 gr. lead bullet
10 grs. black powder

This item was originally developed by the Peters Cartridge Company in co-operation with the J..Stevens Arms Company for their rifles. It was subsequently produced by other manufacturers.

CALIBER 30

(30 Short)
Total length 13/16 in.
1/2 in. copper case, .292 dia.
55 gr. conical, round nose, lead bullet
6 grs. powder

G. R. Stetson is listed in the patent papers as having secured a patent Oct. 31st, 1871, No. 120,403, which reads "Bullet swaged after being placed in shell." There is no record of a patent being issued to him on the 24th, as noted on the box label. The familiar Sharps four-barrel, Stevens single shot and other early day arms used this cartridge.

 50   SHORT   No. 30
  METALLIC CARTRIDGES
  Manufactured by the
WINCHESTER REPEATING ARMS
    CO.

New Haven, Conn., U. S. A.
Trade Mark   Stetson's
    H   Patent
                           Oct. 24th, 1871

CALIBER 31

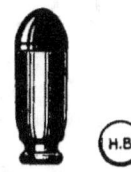

(31 Unknown)
Total length $2^{9}/_{32}$ in.
$2^{1}/_{32}$ in. reverse taper, brass case, .283 dia. above head
Conical lead bullet

Another of the many unknown cartridges which collectors encounter. This one was brought back from Lima, Peru, although the head stamp H. B. is of the well known Houllier & Blanchard of Paris. No doubt a front loader . . . but unknown.

CALIBER 31

(31 Eley Dished Base)
Total length $15/_{16}$ in.
$1/2$ in. copper case, .318 dia.
100 gr. conical, lead bullet
4 grs. powder

From an original box marked only:

50 ELEY DISHED BASE RIMFIRE
    CARTRIDGES
    Powder 4 grains
    Bullet 100 grains
Mfg. by
    ELEY BROS. LTD., LONDON
      "For the Patentee"

CALIBER 32

(32 Extra Short)
Total length $9/_{16}$ in.

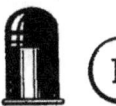

$11/_{32}$ in. copper case
60 gr. cylindrical, round nose bullet
$5\frac{1}{2}$ grs. black powder

Along in the 90's a freakish gun known as the palm pistol was placed on the American market. It used the small cartridge illustrated here, which is sometimes referred to as ".32 Protector." Hartley & Graham's 1897 catalog lists this as being for use in the Remington Magazine Pistol and the New "Protector" Pistol . . . (Chicago Firearms Co.). (See plates 134, 140, *Hand Cannon to Automatic*)

CALIBER 32

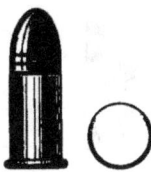

(32 Short)
Total length $15/_{16}$ in.
$17/_{32}$ in. copper case, .317 dia.
80 gr. conical lead bullet
9 grs. black powder

The Smith & Wesson patent of April 17, 1860, No. 27,933 was for . . . "Fulminate in rim, surrounding perforated base wad." As noted on the label, these cartridges were produced by UMC under this patent.

50 No. 2   or   32-100 SHORT
    PISTOL CARTRIDGES
    Manufactured by
UNION METALLIC CARTRIDGE CO.
  Under Smith & Wesson Pat.
      April 17, 1860
     Bridgeport, Conn.

CALIBER 32

(32 Long Rifle)
Total length $1^{3}/_{16}$ in.

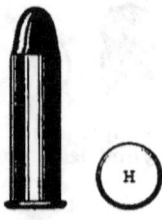

$2^{9}/_{32}$ in. copper case, .317 dia.
Conical lead bullet

An obsolete number, this is a cross between the .32 long and the .32 extra long. It was produced by Winchester and is to be found on some of their old cartridge boards.

CALIBER 32

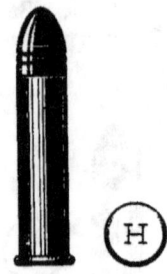

(32 Extra Long)
Total length 1½ in.
1⅛ in. copper case, .317 dia.
90 gr. conical, lead bullet
20 grs. powder
Manufactured by Winchester

This extra long was used in Remington, Stevens, Ballard, Wesson and other rifles. The Remington No. 1 Sporting rifle using it was one of the first of the rolling-block actions to appear in any quantity. Appearing on the market around 1869, it enjoyed a twenty-year popularity.

CALIBER 350

(.350 German R. F. Revolver)
Total length 1⁵/₃₂ in.
²⁵/₃₂ in. copper case
   Dia. at head .343
   Dia. at mouth .343
Conical, lead bullet

For some reason referred to as the

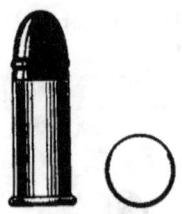

system "Sharp's," this cartridge is one of the early German rim fires. It was used in the cheaper grade of rifles. The one illustrated was made by George Egestorff, Linden, Hanover, Germany.

CALIBER 35

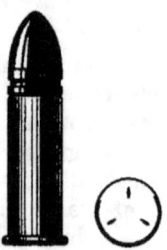

(35 Allen Rifle No. 62)
Total length 1⅜ in.
⅞ in. copper case, .343 dia.
Conical, lead bullet

The label reads as follows:

50 - No. 62
Metallic Waterproof Cartridges
for
Allen's Breech Loading Rifle
Manufactured by
Ethan Allen & Co.
Worcester, Mass.

The Allen breech-loading, single shot rifle was the one in which the breech was lowered straight down by swinging down the trigger guard.

CALIBER 9 mm.

(9 mm. ball)
Total length ⁹/₁₆ in.
⅜ in. copper case, .344 dia.

Spherical lead ball

This number was developed for use in the European Flobert rifles and pistols. They are sometimes called Gallery Cap, since they are primarily for target shooting. The acorn headstamp is the mark of Gustave Genschow of Germany.

CALIBER 38

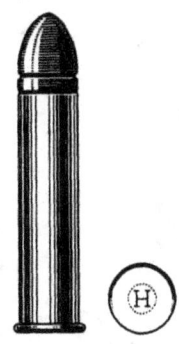

(38 Extra Long)
Total length 1¹⁵⁄₁₆ in.
1⁷⁄₁₆ in. copper case, .377 dia.
148 gr. cylindrical bullet
38 grs. powder

Adapted to the Remington, F. Wesson, Robinson, Ballard, Howard and other rifles. The raised H in a circle denotes an earlier manufacture than the plain stamped letter H.

CALIBER 41

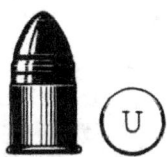

(41 Short)
Total length ⅞ in.
⁷⁄₁₆ in. copper case, .406 dia.
130 gr. conical, lead bullet
10 grs. black powder

A popular class of hand guns which appeal to collectors is the derringer . . . small pocket gun of large bore. For the most part they were carried as auxiliary guns . . . not only by gamblers, but law enforcement men as well. Many citizens were armed with these potent persuaders of other days. A typical label reads:

    50 .41 Cal. SHORT
    RIM FIRE CARTRIDGES
    Manufactured by
The Union Metallic Cartridge Co.
    Bridgeport, Conn. U. S. A.
Swaged   Bullets

(See plates, 129, 135, 162, 163, 164, 165 and 166, *Hand Cannon to Automatic*)

CALIBER 41

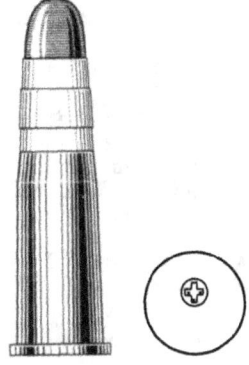

(41 Swiss)
Total length 2³⁄₁₆ in.
1⁷⁄₁₆ in. copper, bottlenecked case
    Dia. at head .539
    Dia. at mouth over paper .449
313 gr. cylindrical, round nose, paper patched, lead bullet
61 grs. black powder

The .41 Swiss cartridge was used in the Swiss Bolt Action 12-shot tubular magazine rifle. This arm was the official Swiss Army rifle from 1869 to 1882. It is one of the very early bolt action magazine rifles for military use.

CALIBER 42

(42 Forehand & Wadsworth)
Total length 1⁷⁄₁₆ in.

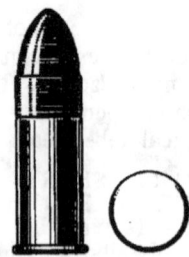

1 3/16 in. copper case, .418 dia.
220 gr. cylindrical, pointed, lead bullet
20 grs. black powder

The descriptive box label reads:

50
No. 64   Cal. 42
Rim Fire Cartridges
with
Swaged Bullets
for the
FOREHAND & WADSWORTH
RIFLE
Manufactured by the
UNION METALLIC CARTRIDGE CO.
Bridgeport, Conn.

Made Expressly For
FOREHAND & WADSWORTH'S
NO. 64
Calibre .42 Rifle

CALIBER 44

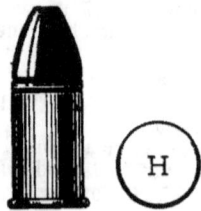

(44 Short)
Total length 1 3/16 in.
1 1/16 in. copper case, .438 dia.
200 gr. conical, flat nose, lead bullet
21 grs. black powder

In addition to other arms, this cartridge was used in the single shot Hammond Bulldog . . . a pocket arm with an unusual side swing breech block.

Trade Mark   44/100
  H
50 CARTRIDGES
Stetson's Patent Oct. 24th, 1871
SHORT
For Pistols
Manufactured by the
Winchester Repeating Arms Co.
New Haven, Conn.

Stetson's patent provided for the bullet being swaged after being placed in the case. (See plate 122, *Hand Cannon to Automatic*)

CALIBER 44

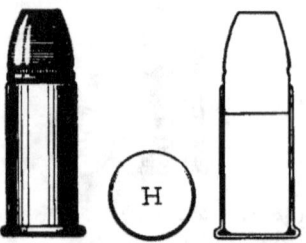

(44 Henry Flat)
Total length 1 11/32 in.
29/32 in. copper case, .445 dia.
200 gr. cylindrical, flat top, lead bullet
26 grs. black powder

Comes now one of the historical cartridges. It was F. Tyler Henry in the late 1850's who was the first to recognize the possibilities of developing the then new rim fire type of cartridges to larger calibers. From 1860 to 1866 ten thousand 12-shot repeating Henry rifles were manufactured. Two regiments of Maj. Gen. Dodge's command were armed with these rifles during their march through Georgia. Many Henry rifles found their way to the West during those early days. Daddy of the famous Winchesters, they hold an important place in arms history. A similar cartridge was made with a pointed lead bullet instead of the flat nose.

CALIBER 44

(44 Extra Long)
Total length 1¹¹⁄₁₆ in.
1⅛ in. copper case, .440 dia.
218 cylindrical, round nose, lead bullet
46 grs. black powder

Developed for the Ballard single-shot rifle, which was first advertised in *Leslie's Weekly* from March 29, 1862, to November, 1862. It was also advertised in *Harper's Weekly* about this same time.

CALIBER 46

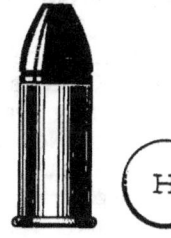

(46 Short)
Total length 1⁵⁄₁₆ in.
²⁷⁄₃₂ in. copper case, .458 dia.
230 gr. conical, flat nose, lead bullet
26 gr. black powder

Used in the Remington Army single-action and other revolvers. The raised "H" on the headstamp is of the earlier Winchester manufacture.

CALIBER 46

(46 Long)
Total length 1¹³⁄₁₆ in.
1¼ in. copper case, .458 dia.
300 gr. cylindrical, flat nose,

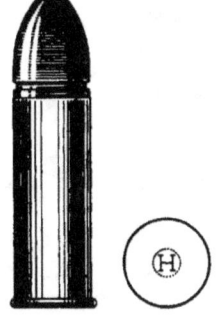

lead bullet
40 grs. black powder

An early manufacture by Winchester. Among the arms in which it was used was the Remington "Split Breech" B. L. carbine. On Jan. 19. 1865, Samuel Norris, an agent of Remington, secured a contract from the War Dept. for 5000 of these arms.

CALIBER 46

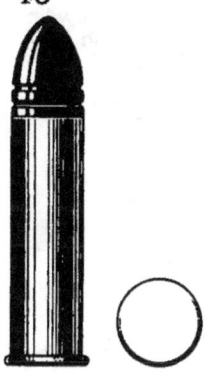

(46 Extra Long)
Total length 2¹⁄₁₆ in.
1¹⁵⁄₃₂ in. copper case
305 conical, pointed, lead bullet
57 grs. black powder

Used in Ballard and other early day rifles.

CALIBER 46

(56-46 Spencer)
Total length 1²¹⁄₃₂ in.
1¼₀ in. copper, bottlenecked case
310 gr. cylindrical, flat nose,

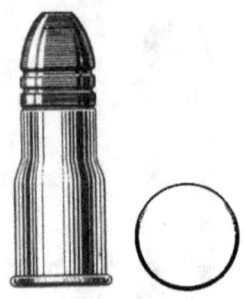

lead bullet
48 grs. black powder

This sporting cartridge was used in the Spencer Repeating Sporting Rifle, cal. 44. Since 1867 the cartridge has been known as the 56-46 Spencer. The 56 refers to the diameter of the case, just above the rim. A description from a Spencer catalog of 1867 reads:

>SPORTING CARTRIDGE
>No. 56 - Caliber 46/100
>Weight of Powder 48 grs.
>Lead 310
>For Rifles of 44/100 caliber

CALIBER 50

(50 Remington Navy)
Total length $1\tfrac{9}{32}$ in.
$\tfrac{27}{32}$ in. copper case
   Dia. at head .559
   Dia. at mouth .535
290 gr. conical, flat nose, lead bullet
23 grs. black powder

One of the early metallic self-exploding cartridges which was in general use for military purposes was the one here illustrated. It was developed for the 50 cal. Remington single-shot navy pistol of 1866.

CALIBER 50

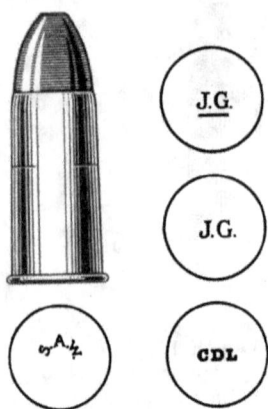

(56-50 Spencer)
Total length $1\tfrac{5}{8}$ in.
$1\tfrac{5}{32}$ in. copper tapering case
   Dia. at head .556
   Dia. at mouth .543
350 gr. conical, flat nose, lead bullet
45 grs. black powder

The carbine (Spencer Repeating Carbine, Model 1865), using these cartridges came out too late for use in the Civil War. For most part they were issued to troops in the West for Indian fighting. Custer's famous Seventh U. S. Cavalry was armed with these carbines on Nov. 27, 1868, at the Battle of Washita.

Illustrated with the cartridge are four different headstamps . . . original wording from two different labels follows:

>FORTY-TWO
>METALLIC CARTRIDGES
>for
>Spencer Carbine
>Cal. 50,   Model 1865
>Manufactured by
>SAGE AMMUNITION WORKS
>Middletown, Conn.

>FORTY-TWO
>SMITH & WESSON'S
>Patent Metallic
>PRIMED CARTRIDGES
>Patented April 17, 1860
>For Spencer Carbine—Calibre .50

Model 1865
Manufactured by C. D. LEET
Springfield, Mass.

CALIBER 50

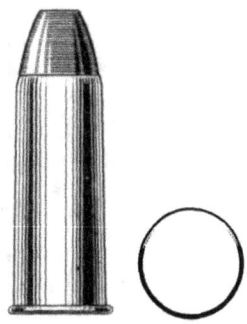

(50 Peabody Musket)
Total length 1¹³⁄₁₆ in.
1⁷⁄₁₆ in. copper tapering case
  Dia. at head .563
  Dia. at mouth .535
320 gr. conical, flat nose, lead bullet
45 grs. black powder

Used in the 50-60 Peabody Musket. In 1865 Canada ordered some 3,000 of these arms.

CALIBER 52

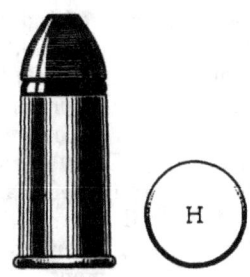

(56-52 Spencer)
Tapered case
Total length 1½ in.
1¹⁄₃₂ in. copper case
  Dia. at head .559
  Dia. at mouth .530
380 gr. conical, lead bullet
48 grs. black powder

(56-52 Spencer)
Bottlenecked case
Total length 1⅝ in.
1¹⁄₃₂ in. copper case

Dia. at head .559
Dia. at mouth .533
385 gr. conical lead bullet
48 gr. black powder

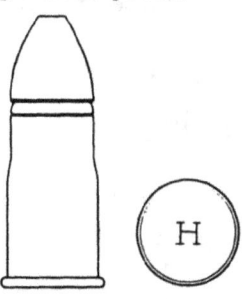

Seven-thousand Spencer repeating rifles were delivered to the War Department during the first six months of 1863. The bottle-necked variety of cartridge was the type used in the Spencer rifles at the battle of Gettysburg. The Spencer was one of the first repeating arms to use metallic cartridges which contained their own priming. Many of these Spencer repeating arms found their way into the West following the Civil War. They were used by many of the early day pioneers for hunting buffalo, deer and such large game. From an old 1867 Spencer catalog is this description of the 56-52 bottleneck:

ARMY CARTRIDGE
No. 56 - Calibre 52/100
Weight of powder 48
Lead 385 grs.
For Rifles of 50/100 calibre

CALIBER 52

(52-70 Sharps)
Total length 2⁵⁄₃₂ in.
1⁷⁄₁₆ in. copper case
  Dia. at head .560
  Dia. at mouth .541
405 gr. cylindrical, lead bullet
70 grs. black powder

Used in the Sharps B. L. military rifle. This rifle is thought to be an alteration of the percussion arms using the .52 linen cartridge to take the .52 rim fire ammunition. It would be logical to assume that the trade of 1867 would be calling for guns using the new type of cartridges.

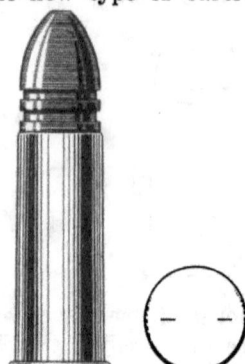

CALIBER 54

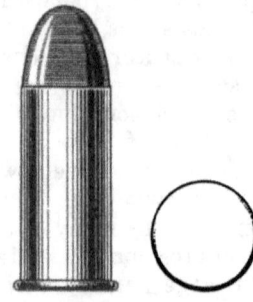

(54 Ballard)
Total length 1²³⁄₃₂ in.
1⁷⁄₃₂ in. copper case
    Dia. at head .577
Conical lead bullet .550 dia.
Black powder

One of the scarce items today is this rim fire Ballard. It was designed for use in the .54 Caliber Ballard Military Rifle and Carbine. Since the guns were only in use for a couple of years (1862-63) the cartridge became an obsolete number quite soon after its introduction.

CALIBER 56

(56-56 Spencer)
Total length 1¹⁷⁄₃₂ in.
⁷⁄₈ in. copper case, .560 dia.
350 gr. conical, pointed, lead

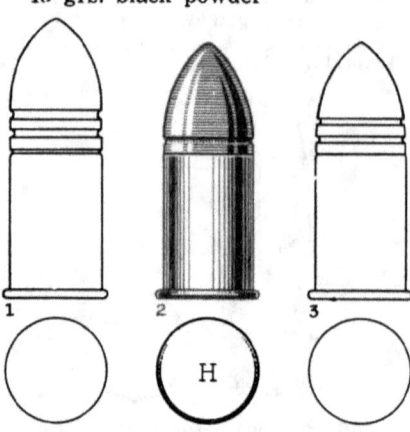

bullet
45 grs. black powder

Illustrated are three of the Spencer 56-56 carbine cartridges. No. 1, total length 1¹¹⁄₁₆ in., is what is known as the navy and infantry size. No. 2 is the later and most common of the Spencer cartridges. No. 3 is but another type of bullet with two cannelures.

Many empty cases of this cartridge are to be found around the location of old forts and out on the plains even today . . . relics of Indian fights and buffalo hunts. Frankford Arsenal produced around 50,000 of these in 1864 and 1865. The priming, using the Sharps mixture, consisted of 6 parts, by weight, of mealed powder—3 of fulminate and 3 of glass. During the Civil War 58,238,924 Spencer cartridges were purchased by the War Department.

CALIBER 58

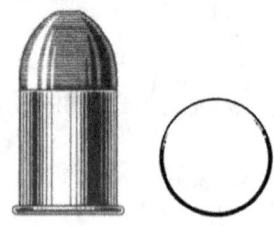

(58 Joslyn Carbine)

Total length 1 7/32 in.
3/4 in. copper case
    Dia. at head .615
    Dia. at mouth .611
380 gr. conical, lead bullet
44 grs. black powder

Used in the Joslyn rim fire carbine, patents of Oct. 8, 1861, and June 24, 1862. The government purchased 515,416 of these cartridges to use with the 11,261 carbines purchased from the Joslyn Fire Arms Co., Stonington, Conn.

A label from an original box reads as follows:

20   .58 Cal. JOSLYN
    RIM FIRE CARTRIDGES
    Joslyn   Carbine
    Manufactured by the
UNION METALLIC CARTRIDGE CO.
    Bridgeport, Conn., U. S. A.

CALIBER 58

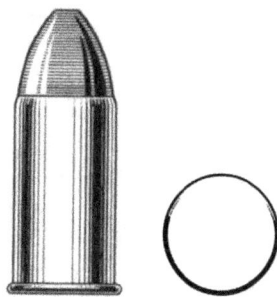

(58 Miller)
Total length 1 11/16 in.
1 3/16 in. copper case
    Dia. at head .631
    Dia. at mouth .621
Conical, lead bullet

At the close of the Civil War plans were undertaken to alter the muzzle-loading Springfield muskets to breech-loading cartridge arms. Cartridges such as this were used in these early alterations. This particular cartridge was used in the Miller swinging block rifle, an experimental of 1867. A postmaster of Meriden, Conn., Miller, the inventor of the Miller alteration, gained prominence when he sued Gen. Benton of the Springfield Armory for infringing upon his patent. He was awarded $20,000 by an act of Congress in 1880 for royalty on the use of his ejector.

CALIBER 58

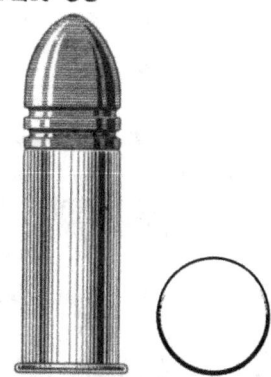

(58 Gatling Gun)
Total length 2 5/32 in.
1 11/32 in. copper, straight case,
    .615 dia.
575 gr. cylindrical lead bullet
70 grs. powder

Dr. Richard Gatling of Chicago in 1862 secured a patent on what is possibly the first machine gun—insofar as the cartridges were fed into the chambers, detonated and extracted by actual machinery operation. In 1866 the gun was altered to take a rim fire cartridge. These guns were manufactured by Colt up until around 1910. Although this cartridge was used in a light field piece rather than a small arm, it is included in this digest as representing one of the first type machine gun cartridges.

CALIBER 58

(58 Mont Storm)
Total length 1 29/32 in.
1 3/16 in. copper case
    Dia. at head .650
    Dia. at mouth .620
480 gr. conical, lead bullet,
    .607 dia.
60 grs. black powder

At one time this cartridge was re-

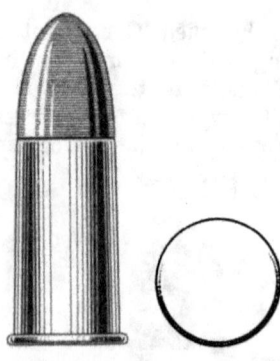

garded among collectors as being of .61 caliber. More recently, however, it has been identified as the .58 Mont Storm. It was used in the Mont Storm conversion of the 1841 U. S. musket. In September, 1858, the government purchased the right to alter 2,000 muzzle-loading muskets to breechloaders under Storm's plan. Only a hundred or so were completed by June 30, 1860.

center fire
PART ONE

INTERNAL PRIMED

EXTERNAL PRIMED

CENTER FIRE PART 1

This section deals with one of the most prolific periods in the entire era of cartridge development. Greater strides were made between 1860 and 1875 than any other like period in history.

Taking the lead in experimenting with early center fire ammunition was the Frankford Arsenal. It was here that many of the early types of center fire underwent rigid tests.

One of the first metallic cartridges to be experimented with at Frankford was the Morse in 1858. It made use of an interior anvil attached to the walls of the case. The separate head was of gum elastic and contained a primer similar to a percussion cap. This complete head assembly was pressed into the case so that the primer came into contact with the anvil. Evidently few were made since specimens are rarely met with today.

Many of the first experiments were with inside primed cartridges. That is, the fulminate was held in place inside the case by means of a cup or anvil. To all outward appearances they resembled their rim fire relative. One type, the Martin, made use of a primer pocket formed out of one continuous piece of the metal case.

As the developments progressed, the inside priming gave way to a solid head with a primer pocket and outside primer ... forerunner of the present day cartridge.

In order to give a clearer conception of this interesting period of cartridge history, a large group of cross-sections of Frankford regulation and experimentals are included. They are redrawn from *Ordnance Memoranda No. 14, Metallic Cartridges*. And, since this is a seldom encountered book among collectors, the actual description of the cartridges are also given verbatim.

ORDNANCE MEMORANDA NO. 14
METALLIC CARTRIDGES
(Regulation and Experimental)
as
Manufactured and Tested at the Frankford Arsenal
Philadelphia, Pa.

Prepared Under the Direction of the Chief of Ordnance by
MAJOR T. J. TREADWELL, Ordnance Department
Commanding Frankford Arsenal

WASHINGTON
Government Printing Office
1873

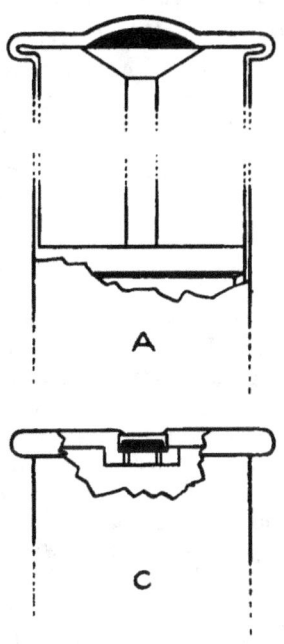

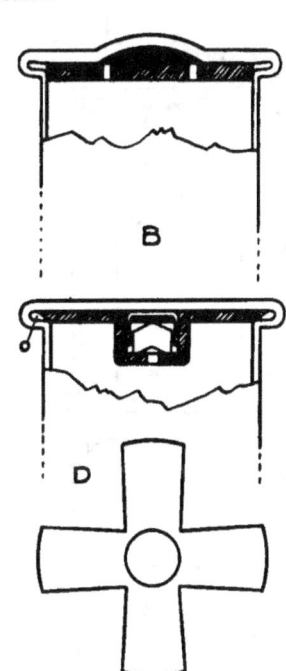

Previous to the year 1866 experience in the manufacture of metallic cartridges at this Arsenal was limited to making a few of the Morse, Burnsides, Maynards and rim fire cartridges for experimental purposes. In the early part of 1864 Col. Laidley com'dg., Special Machinery, (Draw Presses) was introduced preparatory to making cartridges. In 1865, Col. Benét com'dg., a few experimental

"Gatling" 1 in. calibre rim fire cartridges were made to test the Gatling Gun. In 1866, it being evident that the rim fire would be superseded by centre-fire, considerable attention was given to the production of a reliable centre-fire cartridge. Samples of the first attempts to make centre-fire cartridges are shown at A, B, C, D: the case at C has a small cap containing the composition set on the bottom of case without anvil and has the metal pressed over on the cap to hold it in place; it was difficult to make a gas check with it. The case at D was an attempt to make an inside primer by a blank punched out like a star and then formed into a cup holding the anvil and cap with wings which were forced into place by stretching out the wings and securing them in the flange at 0. 50 were fired without failure.

COL. LAIDLEY'S PATENT 1865
Experimental

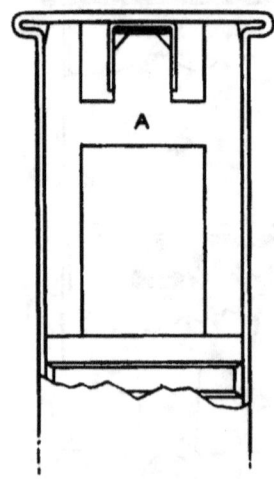

In this invention the cartridge is primed by means of an ordinary percussion cap supported by an anvil or stem resting against the base of ball and kept in its central position by lateral projections or wings in contact with the sides of the cartridge case but not attached to either the base or case.

Claim . . . The combination of an anvil A with a cartridge case of a primed cartridge, the said anvil, not attached to the case, and receiving the percussion cap or priming on one end, the other end resting firmly against the projectile, and of such shape that when inserted it takes a central position and cannot be blown out of the case, which is tapered or contracted at its forward end; the whole as above described and for the purpose specified.

A number of the above were made for experimental purposes at the National Armory, Springfield, Mass. The anvil was punched from thin sheet iron, capped and inserted in the case which was prepared at the open end to allow its mouth elasticity in restoring the case to its original shape, and securing the anvil.

NOVELTY CENTRE PRIMED
Experimental 1866

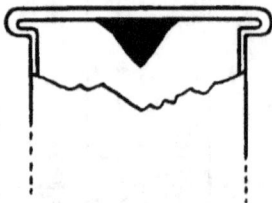

Remarks: Two Frenchmen appeared at this Arsenal about April, 1866, during the administration of Col. S. V. Benet, with a secret composition, proposing its use for centre priming. The composition in a wet state was deposited centrally on the bottom of the case, adhering sufficiently to the metal when dry and surrounded by compressed gunpowder to ignite with a blow. Twenty cartridges were fired with one failure, the composition was very sensitive and great care was necessary in loading to avoid explosions. It would be liable to become detached in transportation and in service.

BENTON'S CUP-REINFORCE
U. S. Armory, Springfield, Mass.
Frankford Arsenal, Pa.
1867

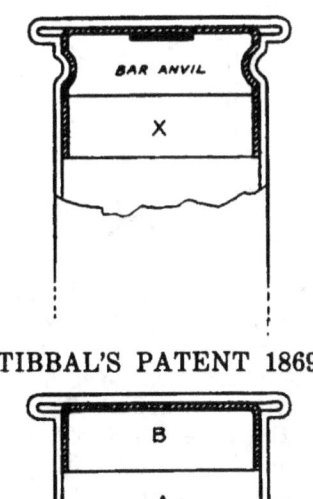

TIBBAL'S PATENT 1869

The bar-anvil cartridge was first made at the National Armory, Springfield, Mass., in 1866 for experimental purposes, by E. H. Martin, (it was invented and submitted to Col. Benton in June, 1866) who was employed at the Armory on cartridge work under the direction of Col. Benton. Its peculiarity at that time was its simplicity and mode of attaching the anvil to the case. It consisted of a copper case and a rigid tinned iron anvil recessed at the centre to hold the percussion composition and grooved in the ends for crimping and securing it in place. Several millions were manufactured at this Arsenal from October, 1866 to March, 1868, when it was superseded by Col. Benét's cup-anvil cartridge.

Several objections were urged against the use of a bar-anvil: it was occasionally thrown into the barrel of the gun when firing: secondly, it was liable to be turned upside down: also crimping the case close under the head drew in the flange producing tension and causing occasional bursting. Col. Benton proposed and applied, in 1867, to the inside of the metallic case, *Fig. X*, a cup reinforce made from thin metal to protect or prevent the gases from reaching the fold; this remedied the last named objective but created another, occasional miss-fires from two thicknesses of metal at point of impact of the firing pin; this may be overcome by opening the case to allow the firing pin to act directly on the inside cup ... It may be here remarked that, so far as can be learned, the above application of a cup as a gas-check to a flanged metallic cartridge case was not especially claimed or patented until 1869, (2 years later than Benton's application of the cup reinforce,) which is as follows: Patent Report, 1869, No. 90,607 to Wm. Tibbals. *Claim* ... The cup or reinforce B, when inserted within the flanged metallic case A, in such a manner as to cover and protect the flange, substantially as described.

BENET'S CENTRE PRIMED EXPERIMENTAL
Frankford Arsenal ... Jan. 1866 Apr.

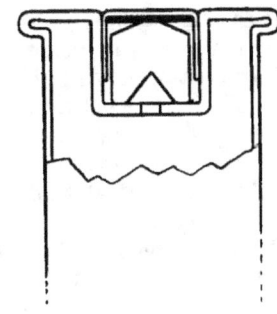

Remarks: The principal feature of this cartridge is the forming of the pocket of one continuous piece of metal. It is believed to have been invented and successfully carried out at the Frankford Arsenal by Col. S. V. Benét, comd'g, in 1866. It is now one of the principal features of Berdan's Cartridge, he having come to the Arsenal and obtained the necessary information, taking with him samples and sizes of tools and afterwards applying it to his cartridge, which previously had a separate cup inserted at the head.

COL. CRISPIN'S Combination Paper and Metal Wrapped Case

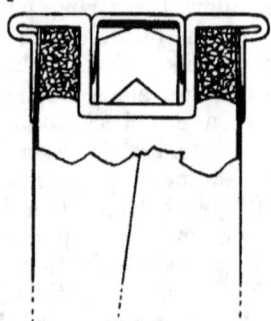

Remarks: In 1867 a number of these cartridges were made at this Arsenal for experiment on the plans of Col. Silas Crispin, Ordnance Dept. as follows . . . A strip of thin sheet brass about ".002 thick was rolled by hand on a roller in connection with a sheet of paper, forming the case of three thicknesses of paper and two of metal, the paper covering the inside and outside of the case, having the metal between. The case was held to a brass head or cap by the friction of a paper wad. A number were fired extracting easily. This mode of attaching is not reliable it being affected by time and atmospheric changes in the loosening of the case from the head.

MARTIN'S CARTRIDGE
1st Patent 1869

Patent No. 88,191 . . . *Claim* —1st: An interior conical shaped pocket or

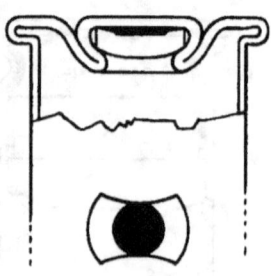

receptacle containing the fulminate and anvil where the wall of said pocket is formed of two thicknesses of metal contiguous to each other substantially as described. 2nd: Turning over the upper part of the conical portion of the reinforcing cup upon and into the pocket or receptacle for the fulminate and anvil forming a gas check substantially as described. Invention patented and the use of it assigned to the United States by Edward Martin employed on experimental cartridge work at the National Armory, Springfield, Mass., Col. J. G. Benton, commanding, 1869. They were first made at the Armory for experimental purposes; during the year 1870 an additional fold under the head was added and patented by the inventor. A number of cartridges of the first patent were made at the Frankford Arsenal in 1871 for the Navy Carbine, Remington pistol and Colt's revolver, cals. .44 and .50. The peculiarity of the cartridge is the forming of an inside pocket from one continuous piece of metal and is performed at two operations.

MARTIN'S CARTRIDGE SERVICE
2nd Patent 1870
Manufactured Frankford Arsenal 1871

Patent No. 111,856—1870 *Claim*— 1st: A metallic shell having the fold C made therein, making the shell of three distinct thicknesses at the juncture of the head with the cylindrical part as set forth. *Claim* . . . 2nd: An annular fillet or corrugation upon the interior of the head of the shell in combination with fold *C*, all constructed substantially in the manner and for the purpose specified.

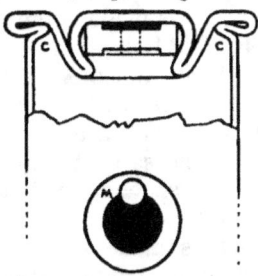

A large number of the Martin cartridges, cal. .50 were made at this Arsenal from May to Dec., 1871. . . . in Oct. a double indented disc-anvil, vented with one vent as at *M*, was suggested by Col. Treadwell and exclusively used thereafter. . . . the peculiarity of the cartridge is the forming of an inside pocket to hold the anvil and the fold *C* of one continuous piece of metal. The object of fold *C* is to give elasticity to the head. The ordinary plain folded head is rigid, and under strain, and may be more or less demoralized by bending, according to the qualities of metal. The fold *C* is supposed to give relief by yielding to the sudden force that would otherwise burst a more than ordinary demoralized or defective plain folded head. (manufacture abandoned at Frankford in December 1871)

DISC-ANVIL
Class 1

As the cup-anvil has no value as a reinforce it was thought the disc-anvil could be substituted on the score of economy, as it saved about 1/3 of metal and one operation, its rigidity and certainty of fire being equal to the cup-anvil. About 30,000 were

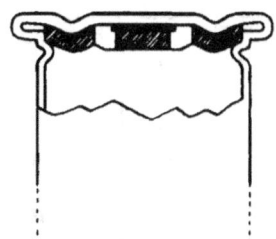

made at this Arsenal in September and October, 1870 for experimental firing and service. At the firing in the gun the expansion of the crimp caused the anvil to become loose and sometimes by the reaction of the gas was driven towards the end of the case . . . it could not get out as the case at the end was smaller than the disc; by the use of a heavy crimp and a gun chamber of minimum size it remained in its place with looseness. In manufacture the same tools and machinery were used as with the cup-anvil; preference is given to the cup-anvil as it is more easily handled in assembling and is not liable to be turned upside down.

CORLISS' FRONT IGNITION EXPERIMENTAL
Class 1

Remarks: The Corliss Needle Cartridge for Springfield B. L. R. M. The needle was attached at the base to a disc crimped into the flange of the case. The point was made to conform exactly with that of the service firing pin. The fulminate was held in a copper capsule in the base of the bullet open to the rear and covered with tin foil. In consequence of the extremely limited longitudinal motion which the needle experiences from the indentation of the cartridge case the bullet was carefully gauged in the loader to bring the fulminate almost in contact with the needle point. Notwithstanding the close gauge employed the first 25 cartridges failed

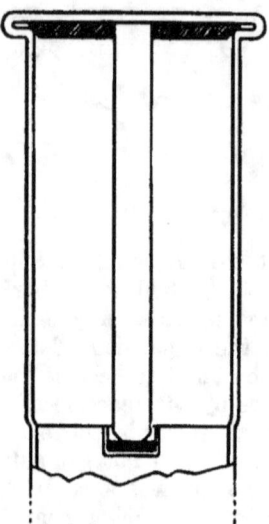

entirely to explode in the gun. On examination the fulminate was found to have been forced into the bullet instead of being crushed by the needle point. 30 more cartridges were made with a needle flat fronted and squared at its apex like the arbor of a watch; which it was thought would effectually explode the fulminate. Out of 18, 5 of them exploded.

MILBANK'S PRIMER and RELOADER
Patented May 1870   Class 1

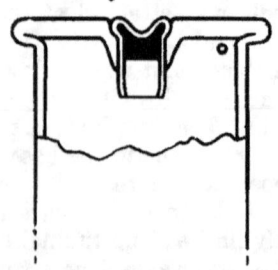

Remarks: A number of the above cartridges, Milbank's Patent, (May, 1870) manufactured by the Winchester Repeating Arms Co., New Haven, Conn. labeled solid head, central fire reloading cartridges, were tested at this Arsenal April 1873.

The head is claimed to be a solid one, which is presumed to mean one folded closely as at O, similar to the Dutch Carbine cartridge. As it does not fulfil the conditions of a solid head it is rated under Class 1. The metal of which the case is made is quite thick, about ".05. The flanges have a greater variation in thickness than any other cartridge tested here, viz: from ".062 to .079 and are variable in thickness in the same flange at different points, showing bad work at an important point.

The primer is made somewhat like a rim fire case, having a recess in the centre holding the priming, partly in the folded rim and at the bottom. The composition is of a dark color resembling that in the Eley cap and is covered by a paper wad, the open end being slightly closed to facilitate its insertion into the pocket of the case.

It was claimed to be superior in the following points: 1st—Certainty of fire. 2nd—Non escape of gas at primer. 3rd—Impervious to moisture. 4th—Facility in reloading.

GAS CHECK REINFORCEMENTS
Class 2

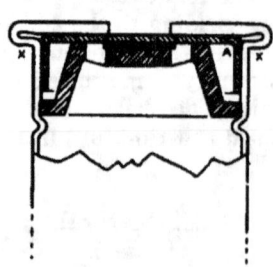

Class 2 consists of cartridges having the flange or head folded on from continuous metal, with additional parts as reinforcements or gas checks, which are intended to prevent the gas from reaching the annular space formed by the bend of metal at flange, as at X. All experience shows the necessity of protecting this weak point of a folded head cartridge, where it is desired to have a case that will at all times prevent the escape of gas from the base, from bursting caused by defective metal or workmanship. Various kinds of material have been tried, such as copper, brass, lead solder, graphite and paper. The essentials of an effectual reinforce or gas check are,—1st: that it must not flow into the annular space at bend transmitting the gas pressure as by the use of soft material as lead, solder, graphite, wax, Vc. 2nd: it must be of such form and construction as will be acted on and expanded quicker than its case or covering. 3rd: it is required to fit tightly against the part to be protected without space to prevent the formation of a volume of gas behind it. The best form is that of a cup, as at A, which is used extensively in hydraulic pumps. . . . About 25 were fired with good results.

BERDAN PATENTS

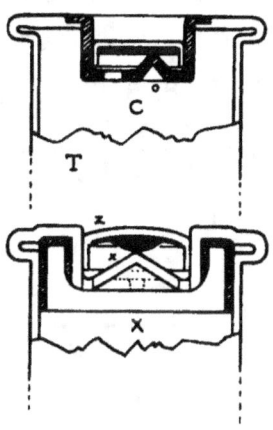

The case, *Fig. C*, having the anvil or teat thrown up from the bottom of cup as at O in combination with a primer or cap is known as the Berdan anvil.—Patent Reports, 1866, No. 53,338.—After seeing the pocket formed of continuous metal, Benets, at this Arsenal, 1866, he very quietly applied his teat anvil to it, soon however changing the teat or projection from side as at O to the centre, *Fig. X*, forming in connection with the Hobbs primer and an interior gas check cup with brass cover a very reliable cartridge. A large number of these cartridges have been made for the Russian Government by whom it is used as a reloader. The throwing up of a portion of the pocket to form the teat or anvil is quite ingenious, but the first method adopted as shown at O *Fig. T*, is believed to be of English origin and has a very different value from the present one used in *Fig. X* as an effective anvil, and the former cannot be well used as a reloader. The cap at Z is known as the Hobbs primer from his patent dated Sept. 14, 1869, No. 94,743;—*claim* . . . a percussion for guns enclosed between varnished surfaces. Its peculiarity is the use of a small quantity of composition rounded in the center and held in place by a varnished tin foil covering, which is attached to the metal of crown by varnish; the cap is of brass and is coated with varnish on the bottom to prevent amalgamation, making altogether a very good primer. About 2,000 of the Berdan centre-fire cartridges have been used here in experimental firing and it is thought to be a first class cartridge.

COL. TREADWELL'S EXPERIMENTS—ATTACHED HEADS
Class 3

The Frankford case, *Fig. A*, was a suggestion arising out of the consideration of the cost and manufacture of wrapped metal cartridges at this Arsenal, November, 1871. It was thought that a brass case without a folded head could be drawn thin and made up with fewer operations and

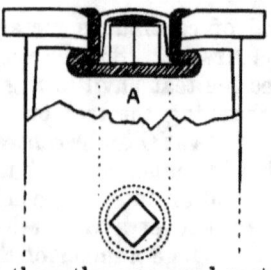

less cost than the wrapped metal and have a strong base, ease of extraction, besides being waterproof. For experiment a number were made and fired with 80 grs. Rifle powder and 300 grs. bullet in a .40 calibre 18 inch twist rifle; the extraction was very easy, the head was a little loose sideways caused by the expansion of case; two cases were reloaded and fired, extracting with freedom; the test was a severe one, proving its strength of base. In experimental firing several Martin cases, with cal. .40 rifle cut through at the extractor seat; for similar experiment it was necessary to reinforce the cup anvil case, using 80 grs. Musket powder, by an additional thin cup as made at Springfield. A late patent dated Sept. 3, 1872 to Milbank, claims the principal feature of the case, *Fig. A*, as follows: the combination of the head and case and the cup serving as a priming cup, as well as a rivet for securing or aiding in holding said parts together.

COL. TREADWELL'S EXPERI-
MENTS—CAST BASES   Class 3
Frankford Arsenal   Feb., 1872

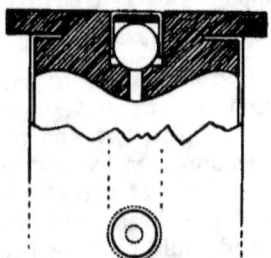

Cast Bases . . . An experiment for casting a head or base on a cartridge case made from paper, foil or drawn metal was made at this Arsenal in February, 1872. It was thought that if a metal or alloy could be found that would flow into mold through small gates and be of such a character as to resist pressure in the gun and stiff enough to insure extraction, that such a case would be economical and could be made in an emergency with less plant and skilled labor than the ordinary drawn case. . . . . A number of alloys made from lead, tin, bismuth, antimony and zinc were made. Considerable difficulty was experienced in casting these sluggish metals. A number of cartridge cases were cast of lead, lead slightly alloyed with antimony, type metal, and type metal with lead. Those of lead were fired in the Springfield gun and their heads spreading to the limit of flange recess and failing to extract, and those from type metal fuzed, breaking off the head in extraction, the metal being too brittle and no arrangement for casting under pressure being at hand the trial was for the present discontinued.

HOTCHKISS'S SOLID HEAD
Patented Experimental   1868

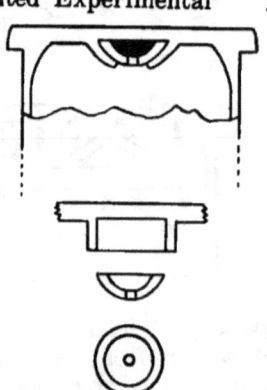

PRINCE'S EXPERIMENTAL

Remarks: The above experimental front ignition tube cartridge, Frank-

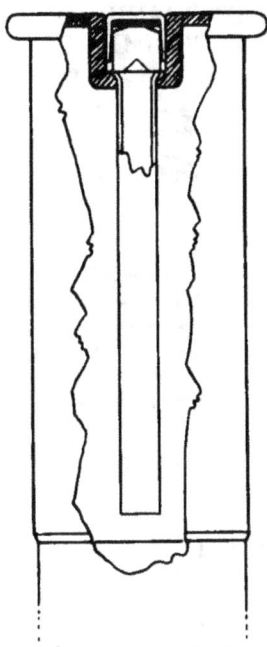

ford Arsenal, suggested by Capt. Prince, consists of a cup case,—a metal tube communicating with the front end of the cartridge reaching to near the head of the bullet—it is designed to be primed with cap and anvil. No experiments have yet been made with this form of cartridge.

BENET'S EXPERIMENTAL
SOLID HEAD
Cup Anvil   1868

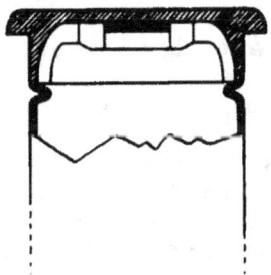

Remarks: The cup anvil for service cartridges was introduced and successfully applied by Col. S. V. Benét, Comd'g. Frankford Arsenal, March, 1868. It was made from tinned sheet iron as an experiment; the crown was indented to hold the priming, with vents for ignition; to keep and support it in place the copper was crimped under its edges. Its superiority over the bar anvil is as follows: —1st: The cup anvil being cylindrical in form furnished more surface at its edges to crimp under, making it more rigid and permanent and greatly increasing the certainty of ignition. 2nd: The crimping and forcing the copper from the sides of the case at a greater distance from the flange considerably lessened the strain at or near the fold, giving a greater strength at head and a smaller percentage of bursting at fold. 3rd: The impossibility of getting the priming upside down, as was sometimes the case with the bar anvil . . .

The tinned cup anvil in damp and salt atmosphere was subject to oxidation thereby destroying the priming composition; it was abandoned August, 1870. Copper was substituted and is now exclusively used for anvils and is not liable to the above defect. Machinery is applied to all the operations in the manufacture, producing a cartridge whose excellence and certainty of fire is not surpassed by any of the various kinds made for the trade. Dimensions of sheet copper for case .028 in. thick, 3.3 in. wide, in strips 36 in. long. Sheet copper for cup anvils .045 in. thick 2.75 in. wide, in strips 25 in. long.

Frankford Arsenal, Penna.
January, 1872

DUTCH MUSKET CARTRIDGE
Class 4

Remarks: The above cartridge . . . was used in connection with guns sent from Holland in experimental of the Calibre Board at this Arsenal. They are designed for reloading shells,— the peculiar feature being the anvil, the leg or handle of which projects through the vent in pocket at O to

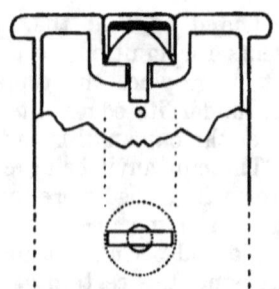

FRONT LUBRICATION
NAVY EXPERIMENT   Class 4
(Farrington's Patent, Dec. 31, 1872)

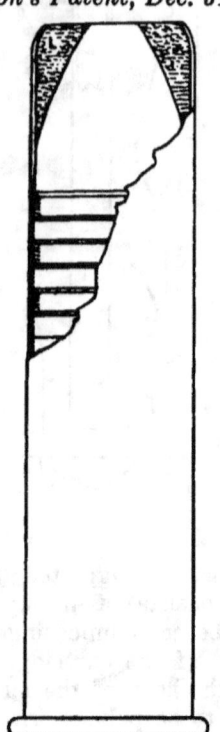

facilitate the extraction of the exploded primer by pressing on the leg from the inside, the cap is easily ejected, a new cap is then inserted on the same anvil and the case reloaded with powder and ball.—The case is of brass, the anvil and primer of copper, the composition of a black color supposed to be in part of mealed powder.

FARRINGTON'S PRIMER SOLID HEAD CARTRIDGE
U. S. Cartridge Co. Lowell, Mass.
Class 4
(Farrington's Improved Patented Dece'r 1872, No. 133,929)

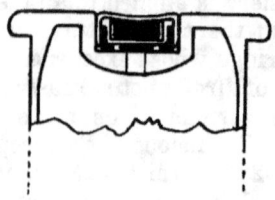

Remarks: At a trial of the above cartridge in comparison with service cup-anvil cartridges some 630 rounds of each were fired, to test the certainty of ignition, in Springfield and Remington rifles with but one failure to explode, (a Lowell cartridge failed) having no fulminate in primer. 100 rounds of each were also fired in the .50 calibre Gatling gun without failure. . . . The solid head case makes a most excellent cartridge and could doubtless be used many times as a reloader if desired.

L .. Lubricant of Japan wax & tallow
Remarks: The above are solid head brass case cartridges differing from those described on preceeding page only in the length of case and front lubrication; the front end of case is reduced on the lubricant to hold it in place.

The following programme was carried out in testing the above cartridges with service, the principal object being to ascertain the amount of fouling deposited after a certain number of rounds had been fired.

1st series 103 rounds of service ammunition, Remington gun, gave fouling........23.9 grains
103 rounds of Lowell ammunition Remington gun, longchamber, gave fouling ....11.2 "

2nd series 100 rounds of service ammunition Springfield gun, gave fouling........14.2 grains

100 rounds of Lowell ammunition
Springfield, long chamber, gave
fouling ....13.3 grains

These two last guns are now fouled by 100 rounds each with its own ammunition and are again fouled by an additional 600 rounds each—in all this last series (3rd) 700 rounds each without cleaning.

3rd series 700 rounds of Lowell ammunition
Springfield long chamber, fouling 21.2 grains
700 rounds of service ammunition
Springfield gun, fouling ....16.0 "

The last series, (as the figures will show) threw no additional light on the subject in question. One Lowell cartridge failed to explode,—cause—had no powder charge. Fired in all 803 rounds Lowell ammunition. Fired in all 803 rounds service ammunition.

RELATIVE PRESSURES CRIMPED vs NOT CRIMPED CASE

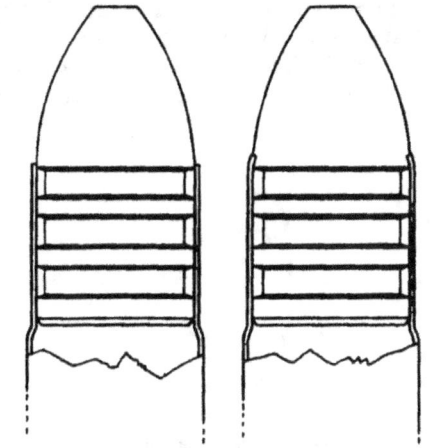

The object of crimping or closing the end of a cartridge case tightly upon the bullet is to insure it against the shock of transportation, the exigencies of service and to exclude moisture. Crimping increases the pressure and consequent strain on the case. With the service ammunition the pressure with crimped case is from two to three thousand pounds per square inch greater than without the crimp and in about the same ratio with calibres .45 and .42. 25 shots of these several calibres, crimped case, gave a mean of 2,300 lbs. per square inch greater than the case not crimped. The velocity is also increased by the use of a crimped case, but not in the same ratio as the pressures; 15 shots of the same calibre gave a mean increased initial velocity of 30 feet per second.

COL. TREADWELL'S EXPERIMENTS - RELATIVE PRESSURES REDUCED vs. STRAIGHT CASE

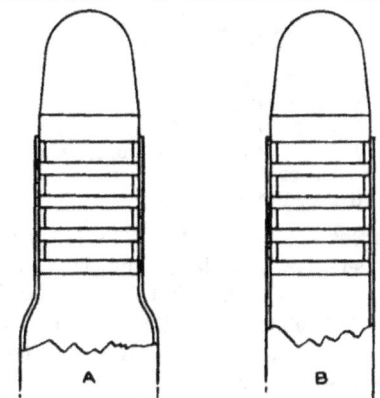

In a reduced or Bottle-shaped case (Service Straight case Cal. .50 reduced for ".40, ".42 and ".45 calibres) as at A, the pressure is greater than in a straight case as at B, both having the same weight of powder and ball.

A mean of 10 shots, Cal. .45, Bottle-shaped Cartridge Case, No. 270 ammunition charge 70 grs. Musket Powder, 400 grs. Bullet, gave a pressure of 18,500 lbs. per square inch.

A mean of 10 shots, Cal. .45 Straight Cartridge Case, No. 272 ammunition, charge 70 grs. Musket Powder, 400 grs. Bullet gave a pressure of 16,300 lbs. per square inch.

A mean of 10 shots, Cal. .42, Bottle-shaped Cartridge Case, No. 271 ammunition, charge 65 grs. Musket Powder, 370 grs. Bullet, gave a pressure of 17,150 lbs. per square inch.

A mean of 10 shots, Cal. .42, Straight Cartridge Case, No. 273 ammunition charge 65 grs. Musket Powder, 365 grs. Bullet, gave a pressure of 16,250 lbs. per square inch.

CALIBER 22

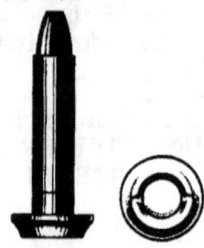

(22 Maynard model 1873)
Total length 1¹¹⁄₃₂ in.
1³⁄₃₂ in. brass case, .234 dia.
45 gr. conical, flat nose, lead bullet
8 grs. black powder

Scarcest of the Model 1873 Maynard calibers, this little fellow was used for both hunting and target work. Various models of the hunting and target rifles of the Maynard 1873 series were chambered for it.

From the best sources available, it would appear that two lengths of cases were made. The small size like the illustration, and a slightly longer (1¼ in.) case.

CALIBER 7.5 mm.

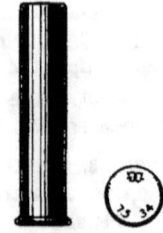

(7.5 Cane Gun)
Total length 1⁵⁄₁₆ in.
Full length brass case

This cartridge, an inside primed type, was used in the European walking sticks, or canes. This particular one bearing the intertwined GG's was manufactured by Gevelot & Gaupillat of France.

CALIBER 307

(.307 Triangular Pistol)
Total length 2⁷⁄₃₂ in.
1⁵⁄₃₂ in. copper case, .307 dia.
Triangular lead bullet.

The distance from the apex to the center of the opposite base was taken as a basis for the caliber of this very odd cartridge. As will be seen from the illustration, the bullet has a slight twist. The primer protrudes about a thirty-second of an inch below the case.

The type of gun in which this cartridge was used is not known. It may have been one of the many experimental "wildcats." However, a bullet mold which casts an identical bullet is known to be in existence.

CALIBER 31

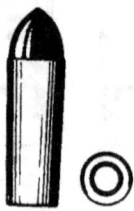

(31 Colt, Thuer's Pat.)
Total length 1³⁄₁₆ in.
²⁷⁄₃₂ in. brass, reverse tapered case
 Dia. at mouth .331
 Dia. at head .312
Conical lead bullet

On Sept. 15, 1868, a patent (No. 82,-258) was granted to F. A. Thuer ... for an improvement in revolving firearms. It was a patent on an alteration of the Colt percussion revolvers, enabling them to fire a tapered metallic cartridge with center fire ignition.

Although this was Colt's answer to the Smith & Wesson controlled (cylinder bored end to end) patent, it did not prove too successful, nor did it enjoy a very long life. An old label reads:

50 $^{31}/_{100}$ in. Caliber
ECONOMIC METALLIC
CARTRIDGES
for
COLT'S new patent REVOLVER
Pocket Size
Manufactured by
COLT'S PATENT FIRE ARMS
MFG. CO.
Hartford, Conn., U. S. A.
The Empty Cartridge Shells Can
Be Reloaded

CALIBER 31

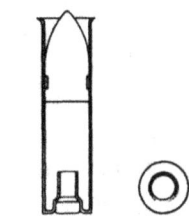

(31 Front Loading—Unknown)
Total length 1$^7/_{32}$ in.
Copper case, .317 dia.
Conical lead bullet

This is an interesting front loader of the early center fire period. It is believed to be equipped with a Milbank primer. The space in the mouth of the case, around the bullet, is filled with wax. It is not known what gun was chambered for this early cartridge.

CALIBER 9 mm.

(9 mm. Cane Gun)
Total length 1$^{17}/_{32}$ in.
Brass case .386 dia.
Spherical lead shot

Another of the cartridges used in the cane guns on the European Continent. It has been said that the owners of estates carried these cane guns as a protection against poachers

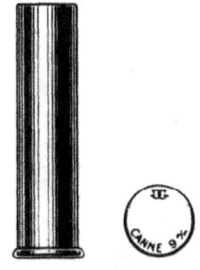

—those who would sneak in to fish or hunt on private grounds.

CALIBER 35

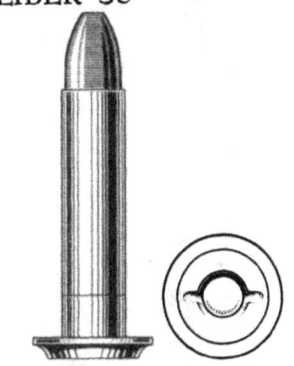

(35-30 Maynard, Model 1873)
Total length 2$^3/_{32}$ in.
1$^5/_8$ in. brass case, .400 dia.
250 gr. conical, flat nose, lead bullet
30 grs. black powder

Edward Maynard secured a patent (No. 135,928) on Feb. 18, 1873, for the altering of his 1865 percussion model arms to ones using a center fire primed cartridge.

These odd cartridges with their large, thick rims, used a Berdan primer. However, they had one central flash hole instead of the usual two or three small holes ordinarily found with Berdan primers.

CALIBER 36

(36 Crispin C. F.)
Total length 1$^1/_{32}$ in.
$^3/_4$ in. brass case
Conical lead bullet

The protruding cap on the head which

contains the fulminate is made of copper. This is but another case of here is the cartridge—but where is the gun that used it?

CALIBER 40

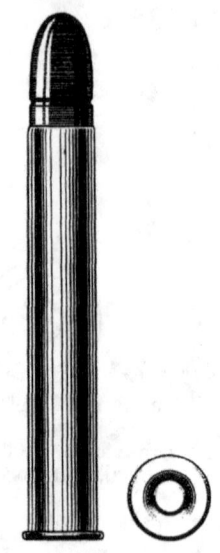

(40 Martin Primed, experimental)
Total length 3³⁄₁₆ in.
2½ in. copper case
  Dia. at rim, .443
  Dia. at mouth, .429
Cylindrical, round nose, lead bullet

One of the early center fire experimentals of the National Armory at Springfield, Mass. This has the Martin primer . . . the one formed out of one continuous piece of the case.

CALIBER 42

(42 Martin Primed, experimental)
Total length 2⅝ in.
2¹⁄₃₂ in. copper necked case

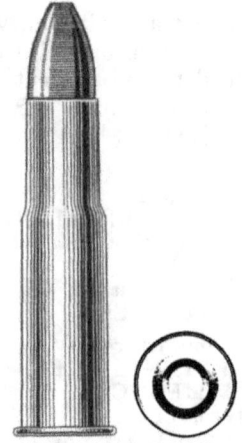

Dia. at rim .555
Dia. at mouth .456
Conical, flat nose, lead bullet

It should be mentioned that during this period of center fire experimentation, many types of cartridges underwent extensive tests at the government arsenals. A collection of these experimentals would be a veritable collection in itself. Most of them, like the one illustrated, are indeed quite rare today.

CALIBER 44

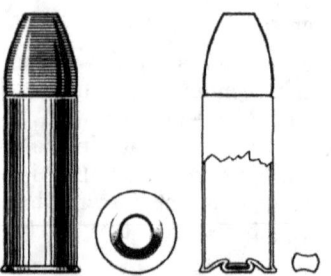

(44 Colt-Remington-Martin Primed)
Total length 1⁹⁄₁₆ in.
1¹⁄₁₆ in. copper straight case, .456 dia.
225 gr. conical, flat nose, lead bullet
30 grs. black powder

The cross-section view illustrates the Martin primer formed out of one continuous piece of the cartridge case. The small piece below the cut-away view is the top view of the tiny anvil.

12 "MARTIN" CARTRIDGES
for
COLT & REMINGTONS ARMY REVOLVERS
Calibre 44
Powder 30   Bullet 225 grains
Frankford Arsenal, Pa.
June 1871
Pat. March 23, 1869 - February 14, 1871

CALIBER 44

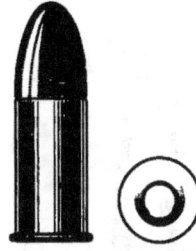

(44 Smith & Wesson - Martin Primed)
Total length 1¹³⁄₃₂ in.
²⁷⁄₃₂ in. copper, straight case, .441 dia.
225 gr. conical lead bullet
25 grs. black powder

March 23, 1869, was the date upon which Edwin Martin secured the patent for . . . "priming cup struck up from base of shell, which may be reinforced." Patent No. 88,191.
A label from an old box reads:

12 "MARTIN" CARTRIDGES
for
SMITH & WESSON'S ARMY REVOLVER
Cal. 44
Powder 25 grains   Bullet 225 grains
National Armory, Springfield, Mass.
April 1871
Patent March 23, 1869—Feb. 14, 1871

CALIBER 44

(44 Colt's Pistol—first alteration)
Total length 1¹¹⁄₃₂ in.
1³⁄₁₆ in. copper, reverse, tapered case
Dia. at mouth .452
Dia. at head .427

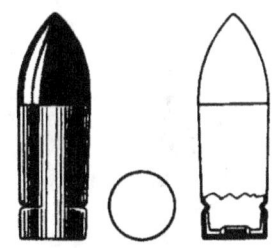

Conical lead bullet .461

This cartridge used the Benet type of inside cup primer. It was purely an experimental cartridge and is quite rare today. *Ordnance Memoranda No. 14*, plate 18, has this to say of it:

"*Remarks:* This cartridge was made as an experiment for use in the first alteration of Colts army revolver. It was inserted into the chamber at front end and held in place by the friction of the bullet and ignited at centre by a firing pin; the friction of the bullet was not at all times sufficient to insure ignition necessarily resulting in miss-fires. A cartridge made with a thin cap and outside priming is said to have worked well."

CALIBER 44

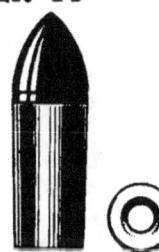

(44 Colt . . . Thuer's Pat.)
Total length 1¹³⁄₃₂ in.
⅞ in. brass, reverse tapered case
Dia. at mouth .461
Dia. at head .431
Conical lead bullet, .461 dia.

Largest size of the front-loading Thuer cartridge with outside center fire primer. This was used in the alteration of the 1860 Colts .44 caliber percussion revolver.

CALIBER 44

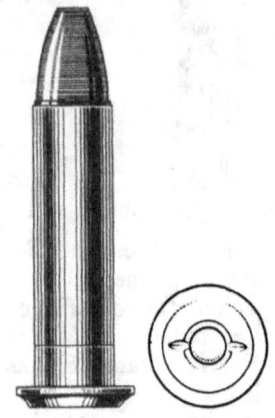

(44-60 Maynard—Model 1873)
Total length 2 7/16 in.
1 27/32 in. brass case, .499 dia.
Conical, flat nose, lead bullet
60 grs. black powder

Used in the 1873 Maynard improved Hunter's rifle No. 11, a rifle which was designed for large game. Once the extractor had started these cartridges out of the chamber after firing, the large head afforded an easy hold for completely removing the case.

CALIBER 44

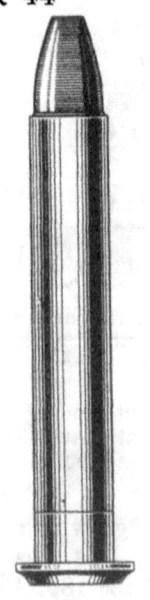

(44-100 Maynard, Model 1873)
Total length 3 15/32 in.
2 7/8 in. brass case, .499 dia.
520 gr. conical, flat nose, lead bullet
100 grs. black powder

Designed for the Model 1873 Maynard Long Range Creedmore Rifle No. 14. An 1885 Maynard catalog says . . . "For hunting and target practice at all ranges, the Maynard more completely supplies the wants of hunters and sportsmen generally than any other rifle in the world."

CALIBER 45

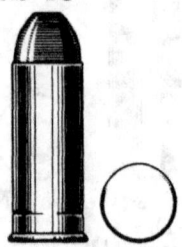

(45 Colt . . . Benét Type)
Total length 1 13/32 in.
1 3/32 in. copper case, .479 dia.
250 gr. conical, flat nose, lead bullet
30 grs. black powder

Used in the Colt Army Revolver Model 1872. Eight thousand of these revolvers were ordered by the War Department in 1873 after thorough tests in 1872. The Benét type of primer is described in the group of Frankford experimentals at the beginning of this section.

CALIBER 45

45 Springfield (inside cup primed)
Total length 2 19/32 in.
2 in. copper, necked case
    Dia. at rim .560
    Dia. at mouth .488
Conical, flat nose, lead bullet

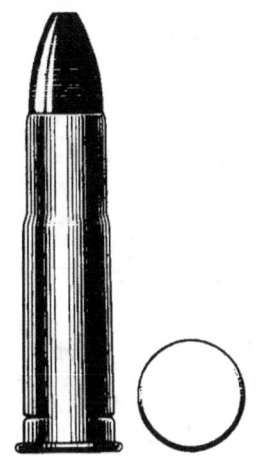

One of the rare Springfield Armory inside cup primed experimentals.

CALIBER 45

(577-450 Martini-Henry)
Total length 3⅛ in.
2½ in. coiled brass, necked case with iron base
    Dia. at head .661
    Dia. at mouth .482
480 gr. cylindrical, paper patched, lead bullet
85 grs. black powder

This unusual appearing cartridge was one of the first English attempts to bottleneck a cartridge. It was nothing more than the 577 Snider case, necked down to hold a 45 caliber bullet.

The cartridge was first tested at the field trials in October of 1870. After extensive test in the Martini-Henry rifle, both cartridge and rifle were adopted in April, 1871, as the official arm and ammunition of the British armed forces. It was in service from 1871 until 1888.

CALIBER 12 mm.

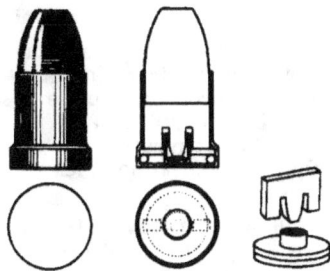

(12 mm. Perrin)
Total length 1⁵⁄₁₆ in.
⁹⁄₁₆ in. copper case, .466 dia.
Conical, flat nose, lead bullet

One of the foreign revolvers used in the Civil War was the Perrin, a French arm manufactured by Perrin & Co. of Paris. The cartridge for this gun is illustrated here in cross-section. As will be noted, this is an inside primed center fire cartridge, not a rim fire. In fact the type of anvil is a great deal like that later patented by Col. Laidley in this country in 1865. The wording, in French, for an original circular label is as follows:

CARTOUCHES POUR REVOLVERS,
SYSTEME PERRIN, Bte.
25
CARTOUCHES
Percussion au Centre
avec Capsule ordre
Inflammation Sous Envelope
System Brevete
a
PARIS
Calibre 12 Mill tres A. Balles

CALIBER 46

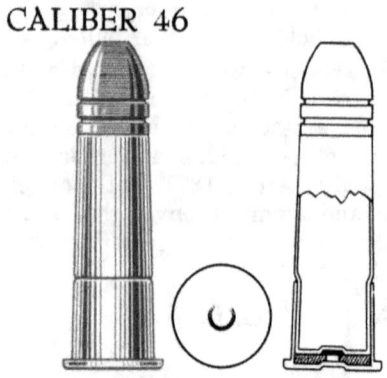

(46 Winchester 2-piece case)
Total length 1 31/32 in.
1 13/32 in. copper, 2-piece case
Dia. at rim .532
Dia. at mouth .463
Cylindrical, flat nose, lead bullet

The following patent (No. 60,814) was granted to O. F. Winchester Jan. 1, 1867 ... "Rubber washer between metal tube and base cup. Fulminate in recess at base of tube ..." This is one of the very rare Winchester experimentals.

Winchester has this to say concerning this cartridge in his application for a patent, "The object of my invention is threefold: first to produce a cartridge that shall be less liable to accidental explosion in handling and also in the magazine of the gun; second, to so strengthen the base or head as to prevent the breaking or pulling off of the flange or head; and third, to produce a central-fire cartridge so constructed as to act as a cushion, for the purpose of relieving the breech-pin more or less from the sudden shock or strain caused by the explosion of the charge."

CALIBER 50

(50 Remington Army)
Benet type—inside primed
Total length 1 1/4 in.
27/32 in. copper, necked case
Dia. at rim .563
Dia. at mouth .532

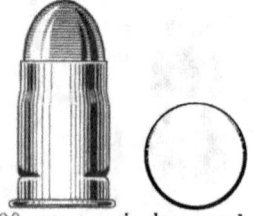

300 gr. conical, round nose, lead bullet
25 grs. black powder

One of the first single shot hand guns used by the army for metallic cartridges was the Remington rolling block pistols of .50 caliber. These were first brought out in 1865.
An old label reads:

12
CENTER PRIMED CARTRIDGES
for
Remington's Army Pistol
Caliber .50
Frankford Arsenal

CALIBER 50

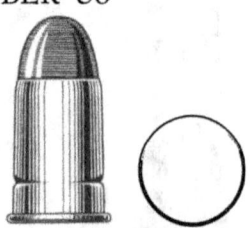

(50 Remington Navy)
Benet type—inside primed
Total length 1 7/32 in.
7/8 in. copper case
Dia. at rim .562
Dia. at mouth .525
300 gr. conical, round nose, lead bullet
25 grs. black powder

Used in the Remington Model 1865 Navy single shot pistols. The cartridge for the Navy pistol differed from that of the Army model. The Navy cartridge has a slight tapered case, while the Army cartridge has a distinct bottleneck.

CALIBER 50

(50 Remington Army)
Martin Primed—Spring-

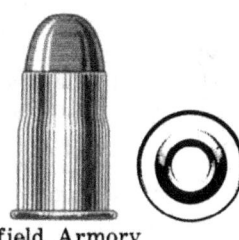

field Armory
Total length 1¼ in.
⅞ in. copper, necked case
  Dia. at rim .565
  Dia. at mouth .535
300 gr. conical, round nose, lead bullet
25 grs. black powder

The Remington single shot .50 caliber Army pistols were used in the service for something like ten years. Many shooters today would like to find these finely balanced old arms for rechambering for present day cartridges.

THE MORSE CARTRIDGES

One of the first of the metallic cartridges to be experimented with was the Morse. So important are these cartridges as the active beginning of center fires, that the continuity of this section, so far as calibers are concerned, is being interrupted to bring these interesting cartridges together.

Two patents concerning the same type of cartridge... No. 20,214 of May 11, 1858, and No. 20,727 of June 29, 1858, were granted to George W. Morse. The first of these reads... "Metal tube has pronged anvil soldered inside. Base closed by cup, which is driven against cap on anvil by hammer in firing." The second provides for a "Cap on anvil in base of metal tube, surrounded by perforated rubber disk."

CALIBER 58
  (58 Morse, early)
  Total length 2¼ in.
  1 9/16 in. brass, brazed seam, tapered case
      Dia. at rim .633
      Dia. at mouth .572

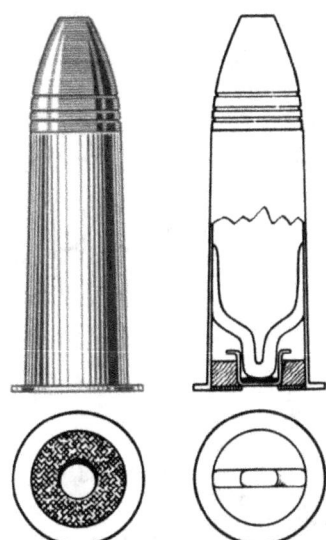

This is the early Morse believed to have been made at Springfield Armory. It has the round wire anvil and the flat solid rim, not found on the later Frankford Arsenal Morse cartridges.

CALIBER 50

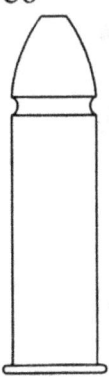

(50 Morse)
Total length 2 5/32 in.
1½ in. brass-straight case with folded rim
    Dia. at rim .546
    Dia. at mouth .543
Conical, flat nose, lead bullet
Flat type anvil as illustrated in detail sketch

CALIBER 54

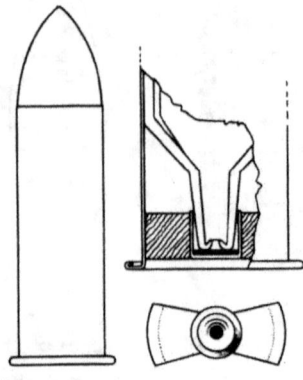

(54 Morse)
Total length 2 5/32 in.
1½ in. copper, straight case, folded rim
    Dia. at rim .574
    Dia. at mouth .513
Conical, pointed lead bullet
Flat type anvil
    Frankford Arsenal Manufacture

CALIBER 58

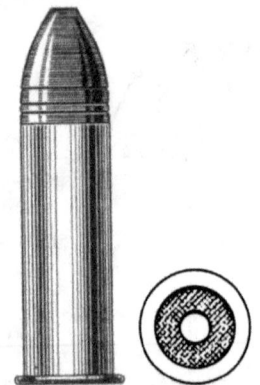

(58 Morse)
Total length 2¼ in.
1½ in. tinned case, with folded rim
    Dia. at rim .576
    Dia. at mouth .576
Conical, flat nose, lead bullet
Flat type anvil
    Frankford Arsenal Manufacture

In addition to the four calibers illustrated, the Morse was made in a .69 caliber and a 16 ga. shot. Several Morse rifles with three barrel combination were made in 1857. A .50 caliber rifle barrel, a .54 caliber carbine barrel, and a 16 ga. shot barrel were supplied as a set. Truly the Morse was a milestone in cartridge history . . . and specimens of it are scarce indeed today.

CALIBER 50

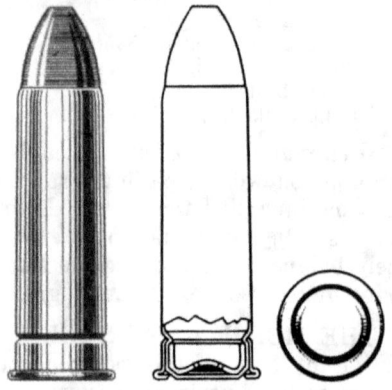

(50 "Moore's Rolled Flange")
Cup Primed
Total length 2¼ in.
1¾ in. copper, tapered case
    Dia. at rim .561
    Dia. at mouth .537
Conical, flat nose, lead bullet

A Frankford Arsenal experimental cartridge. The cross-section shows the "rolled flange," the inside cup and the manner in which the head is made up.

CALIBER 50

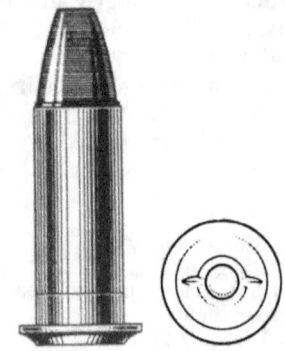

(50-50 Maynard Model 1873)
Total length 2¹/₃₂ in.
1⁷/₁₆ in. brass case, .553
Conical, flat nose, lead bullet
50 grs. black powder

Used in the Maynard improved hunter's rifle, No. 11, for large and dangerous game.

CALIBER 50

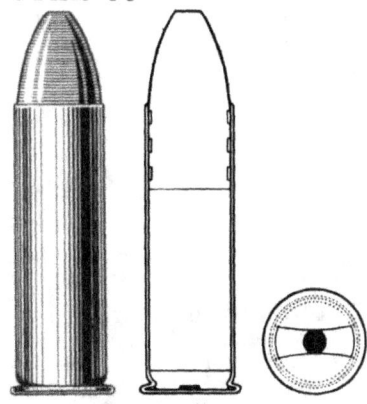

(50-70 Bar Anvil)
Total length 2⁹/₃₂ in.
1¾ in. copper case
450 gr. conical lead bullet
    Dia. at rim .564
    Dia. at mouth .541
70 grs. black powder

Manufactured at Frankford Arsenal between October 1866 and March 1868. This was the cartridge used in the altered Springfield muskets. The small handful of men at the famous Battle of the Wagonboxes were armed with the new breechloaders shooting these cartridges. It was here that 32 pioneers were pitted against 3,000 Indian warriors under Red Cloud . . . and yet won . . . thanks to breech-loading guns . . . and Metallic Center Fire cartridges.

CALIBER 50
(50-70 Benét Inside Primed)
Total length 2⁹/₃₂ in.
1¾ in. copper case
    Dia. at rim .564
    Dia. at mouth .538
450 gr. conical lead bullet
70 grs. black powder

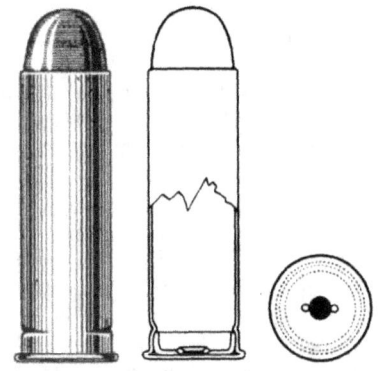

Another of the early center fires used in the muskets which were altered following the Civil War. The label describes it as follows:

20
CENTRE-PRIMED, METALLIC
C A R T R I D G E S
Calibre .50
Charge, 70 grains Musket Powder
Weight of Bullet 450
Frankford Arsenal, March 1871

CALIBER 50

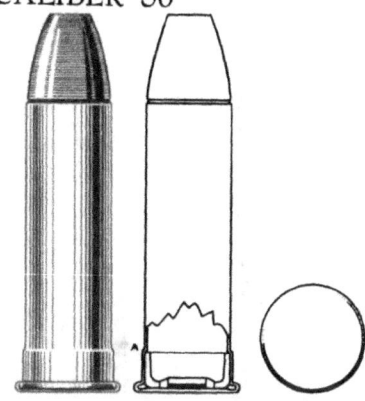

(50-70 Cup Anvil)
Total length 2⁹/₃₂ in.
1¾ in. copper case
    Dia. at head .565
    Dia. at mouth .540
450 gr. conical, flat nose, lead bullet
70 grs. black powder

One of the early center fire government experimentals. It is described in *Ordnance Memo No. 14* as follows:

"As the cup-anvil has a portion of its sides crimped or forced in to give support to the anvil, it was supposed to cause unequal expansion at firing, producing the tightness at extraction complained of in the Remington gun. It was proposed to remedy this by dispensing with the crimp, as it only went part way around the case, and substitute a case reduced part way in size, say .015 in.—which supported the anvil all around its edges at A. A number were fired in the Remington gun, compared with the Martin fold, it worked very well as to extraction with but an occasional tightness by the expansion of the case at the point that held the anvil in place as at A. In manufacture the corners of the reducing die that formed the support of anvil, as at A, wore off rapidly and if not closely watched will make uncertain work, resulting in miss-fires; its value for strength is the same as the ordinary cup-anvil case."

CALIBER 50

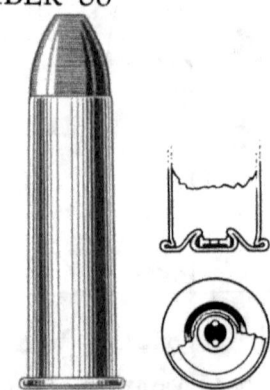

(50-70 Martin Primed)
Total length 2⁹⁄₃₂ in.
1¾ in. copper case
  Dia. at rim .562
  Dia. at mouth .530
Conical, flat nose, lead bullet
70 grs. black powder

The Martin cartridge was made at the National Armory at Springfield under Edwin Martin's patent No. 111,856 of Feb. 14, 1871. It was as follows: "Head of shell folded down to secure anvil. Reinforcing cup may be held by crimping shell above it."

CALIBER 50

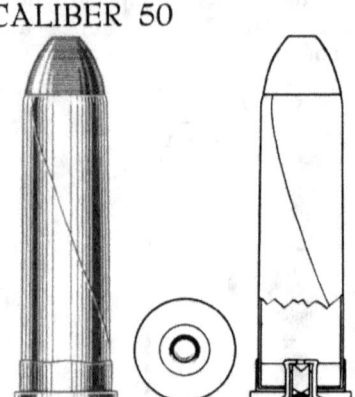

(50-70 Rodman - Crispin)
Total length 2³⁄₁₆ in.
1¹³⁄₁₆ in. coiled brass case
Conical, flat nose, lead bullet
70 grains black powder

About 2,000 of these cartridges were made in January and February, 1872, at Frankford Arsenal. They were made under the Rodman and Crispin Patent (No. 40,988) of Dec. 15, 1863. A patent for a . . . "shell of wrapped sheet metal, with base crimped in and reinforced."

CALIBER 50

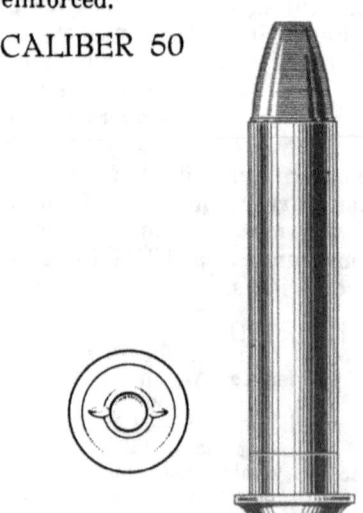

(50-100 Maynard, Model 1873)
Total length 2³¹⁄₃₂ in.
2⅜ in. brass case, .555 dia.
Conical, flat nose, lead bullet
100 grs. black powder

Used in the Maynard improved hunter's rifle, Number 11, one of the big game rifles of the 1870's.

CALIBER 55

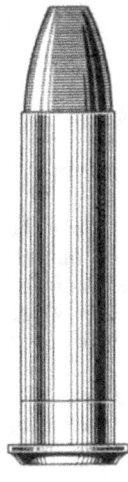

(55-100 Maynard, Model 1873)
Total length 2¾ in.
2⅛ in. brass case, .599
Conical, flat nose, lead bullet
100 grs. black powder

This cartridge is one of the distinct Maynard Model 1873 rarities. Even though the cartridge is listed in the 1885 Maynard catalog, it is found only in a few select collections today. It was another of the large powerful cartridges used in the No. 11 Maynard improved hunter's rifle of 1873.

CALIBER .577

(577 Snider)
Total length 2⁷⁄₁₆ in.
1¹⁵⁄₁₆ in. coiled paper and brass case with iron head
Dia. at rim .655
Dia. at mouth .614
480 gr. conical, lead bullet
70 grs. black powder

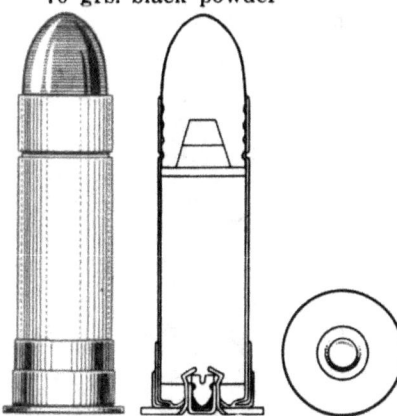

In 1864, after extensive tests, the English army adopted the Snider breech-loading rifle. This was a conversion from the muzzle-loading Enfield musket in use previous to 1864. Originally the Snider was designed to take the Schneider cartridge ... a center fire using a percussion cap with a separate anvil held in a composite head. The head was attached to a cardboard case ... similar to a shotgun shell.

This cartridge was not accepted ... and instead, the developing of a suitable cartridge was turned over to Col. Boxer of the English Ordnance. In 1865 he produced an acceptable cartridge which was adopted and used for many years. It is commonly referred to as the "Boxer" cartridge. The Snider, or Boxer, is composed of several pieces. The cross-section illustrates how the cartridge is made up. Note that the head assembly is attached to the case by the novel manner of having the primer pocket serve as a rivet.

The bullet had a hollow base with a clay plug. Later, for stability at long range, the clay was replaced with a wooden plug.

In 1866 the Snider rifle and the Boxer cartridge were officially adopted as the standard arm and ammunition of the English armed forces.

CALIBER .577

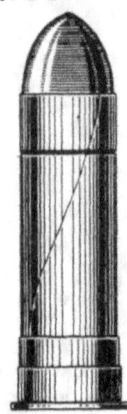

(577 Snider)
Total length 2 7/16 in.
1 15/16 in. coiled, brass case with iron head
    Dia. at rim .660
    Dia. at mouth .605
480 gr. conical, lead bullet
70 grs. black powder

This is the same cartridge as the preceding type with the exception that the case of this one is made entirely of coiled, thin brass.

CALIBER 58

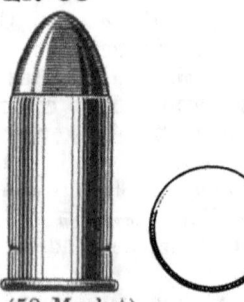

(58 Musket)
Total length 1 11/16 in.
1 3/16 in. copper case
    Dia. at rim .653
    Dia. at mouth .625
Conical lead bullet

Used in the early breech-loading muskets which had been altered from the muzzle-loaders following the Civil War—such as the Allen alteration, etc.

CALIBER 58

(58 Musket . . . Martin Primed)
Total length 1 11/16 in.
1 3/16 in. copper case
Conical lead bullet

This cartridge has the Martin primer pocket formed out of one continuous piece of the case head. It, too, was used in the early transformed muskets. This type used an inside cup reinforce . . . the object of this cup being to reinforce the folded rims and to serve as a gas check.

CALIBER 1 in.

(1 in. Gatling Gun)
Total length 3 7/8 in.
2 25/32 in. copper case
8 oz. conical lead bullet
.75 oz. mortar powder

This cartridge is illustrated merely to show how the inside priming principle of the early center fire types was carried over into the larger arms. The Gatling gun was produced for many years by Colts. Nearly 200 shots a minute could be fired by this early rapid fire gun. An early box label describes this item as follows:

5
CENTER-PRIMED, METALLIC
BALL CARTRIDGES
For Gatling Gun
Calibre 1 inch
Charge, .75 ounce Mortar Powder

Weight of Bullet, 8 ounces
FRANKFORD ARSENAL
1867

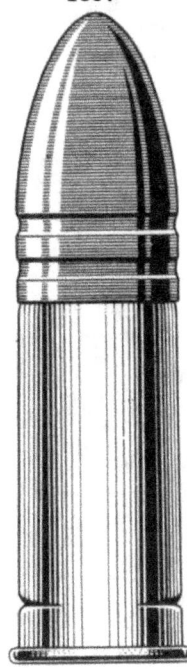

CALIBER 1. in.

(1 in. Nordenfelt)
Total length 4 13/16 in.
3 3/4 in. coiled copper case
3228 gr. bullet
468 grs. powder

As the official British ammunition for the Nordenfelt Machine Gun ... it is said to have undergone experiments for a short time by the U. S. Navy. The Nordenfelt was a stationary gun ... with from two to six barrels placed side by side. The cartridges were placed in a hopper above the barrels, and were fed into the chambers by gravity. The empty shells were extracted by the movement of a hand lever.

This gun was quite popular in Europe during the eighties and early nineties.

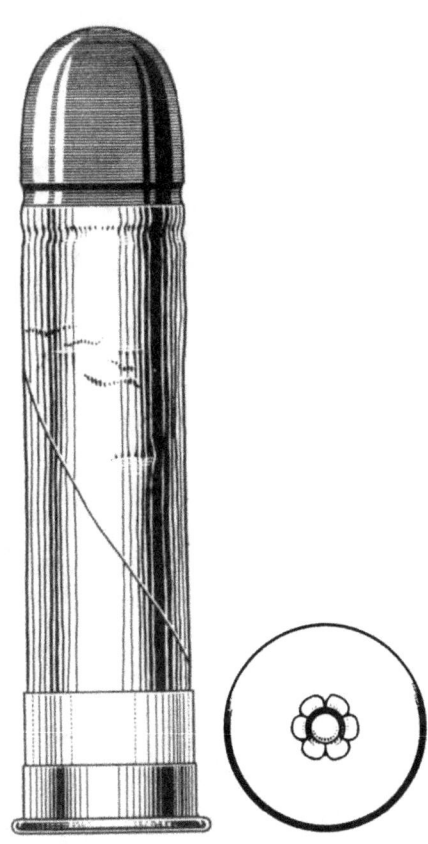

center fire
PART TWO

EXTERNAL PRIMED

CENTER FIRE PART 2

With the expiration of the Smith & Wesson patent, having to do with a cylinder bored through from end to end . . . and the development of the center fire cartridge with outside primer, the field was clear for unlimited research and development. Hundreds of modifications and improvements have been made . . . but basically very little change has been made in recent years in center fire cartridges, compared to the advances made in other periods.

Rimless and semi-rimmed . . . paper patched, hollow point, soft point, metal jacket, smokeless powder and non-corrosive primer, all are a product of this period.

Two types of primers are in use today. The Berdan type in which the anvil is a part of the case itself, and is formed in the primer pocket. The other, and the one in general use in this country is the type in which the primer contains a separate tiny anvil. The latter type is much easier to remove and reload.

The early solid head cases were made with a balloon type primer pocket . . . that is, the pocket extended into the powder chamber of the cartridge. This has largely been replaced by the "Solid Webb," a head in which the primer pocket is formed in a thicker head, leaving the powder chamber flush across its base.

In recent years greater stress has been placed on higher explosive propellants and bullet types than any other phase of cartridge development. Velocity, trajectory and ballistics are bywords among shooters today . . . whereas ball, powder and paper wad were the principal vernacular of a century ago.

CALIBER 2.7 mm.

(2.7 mm. Kolibri Automatic)
Total length 7/16 in. (.431)
11/32 in. brass, rimless case .139 in. dia.
2½ gr. round nose, metal jacket, bullet (.106 dia.)
1½ grs. powder

Smallest automatic cartridge ever developed is this miniature Kolibri. It is used in the tiny German Kolibri automatic pistol. Even though it looks small and unpretentious, it is said that its bullet will penetrate as much as 1½ in. of pine . . . so it is not exactly a plaything.

The cartridge here illustrated was brought back from World War I. One such gun is known to have been taken from a German officer who was using it as a hide away arm in the event of capture.

Boxes from Austria were labeled "2.7 Kolibri."

CALIBER 3 mm.

(3 mm. Kolibri Automatic)
Total length 1 3/32 in. (.430)
5/16 in. brass, rimless case . . . .150 in dia.
Round nose, lead bullet (.119 dia.)

Another of the tiny Kolibri cartridges. These are of a more recent manufacture than the previous one. They were boxed in a white box with a red label which read:

50 Cart Kal 3mm.
Pour Automat Pistolet "KOLIBRI"
"Made in Germany"

These Kolibris are believed to be the smallest center fire cartridge ever manufactured.

CALIBER 4 mm.

(4 mm. Luger Auxiliary)
Total length 11/32 in.
¼ in. copper case
Dia. at head .223
Dia. at mouth .191
7 grs. spherical, lead bullet

A tiny cartridge used in a small caliber barrel within a regular Luger barrel for indoor target practice. This one was manufactured by H. Utendoerffer of Germany.

CALIBER 145

(145 Jones)
Total length 1½ in.
1⅛ in. brass, necked case
Dia. at head .225
Dia. at mouth .171
17 grs. metal jacket, soft point, bullet .149 dia.

The case of this "Wildcat" was necked down from a 5.5 Velo Dog cartridge. It was made to shoot in a rifle which had been built, "lock, stock and barrel." The barrel was turned and bored from an old Model A Ford axle. When American gun enthusiasts get on the beam there is little, if anything, they cannot do.

CALIBER 17
(17 Jones - Scorpion)
Total length 1¾ in.
1⅜ in. brass, necked case
Dia. at head .295
Dia. at mouth .190
Metal jacket, soft point, bullet, .171 dia.

Another Jones "Wildcat." This one necked down from a Hornet case. It was used in a hand made 20 in. barrel on a low wall Winchester action.

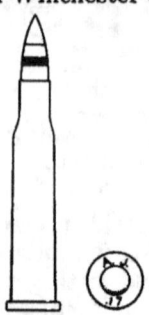

CALIBER 4.25 mm.

4.25 mm. Lilliput
Total length 9/16 in.
1 3/32 in. brass, rimless case, .199
12 gr. metal jacket bullet, .169

August Menz of Suhl, Germany, introduced the small Lilliput automatic pistol which uses this cartridge. It was originally designed for the Austrian Erika pistol. Even though small, it is capable of inflicting a serious injury at very close range.

CALIBER 5 mm.

(5 mm. Clement Automatic)
Total length 1 in.
23/32 in. brass, necked case
Dia. at head .279
Dia. at mouth .221
Metal Jacket bullet, .201

Used in the Belgian Clement automatic pistol and in the Charola-Anitua of Spanish manufacture. The latter was an invention of Theodore Bergmann whose name figures quite prominently in early development of automatic arms.

CALIBER 5.5 mm.

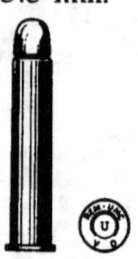

(5.5 mm. Velo Dog)
Total length 1 13/32 in.
1 5/32 in. brass case
Dia. at head .248
Dia. at mouth .243
45 gr. metal jacket bullet

These cartridges were developed for use in small European pocket revolvers — French, Belgian, German, etc. The cylinders on these arms were extra long to accommodate the cartridge. They look quite odd indeed on the small revolvers. Velo Dog cartridges were also made in this country. A label from a box reads:

5.5 mm. Velo Dog
Smokeless
25 Central Fire Cartridges
Metal Cased
Remington UMC
For Velo-Dog Revolver

CALIBER 22

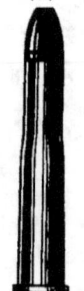

(22 W. C. F.)
Total length 1 23/32 in.
1 13/32 in. brass, necked case
Dia. at head .294
Dia. at mouth .245

45 gr. conical, flat nose, lead bullet
13 grs. black powder

An obsolete cartridge. It was used in the Winchester single shot rifle right at the turn of the century. It occupies a place in this digest because it was the predecessor of the modern .22 Hornet cartridge.
A Winchester catalog of 1885 lists an 1873 model as chambered for this cartridge.

CALIBER 22

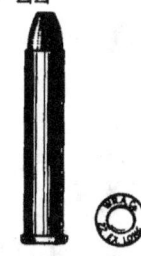

(22 Extra Long Maynard)
Total length $1^{13}/_{32}$ in.
$1^{5}/_{32}$ in. brass straight case, .252 dia.
45 gr. conical, flat nose, lead bullet
8 grs. black powder

Used in the Maynard gallery and Maynard small game hunting rifles, models of 1882. These cartridges were originally primed with the tiny No. 0 primer.

CALIBER 22

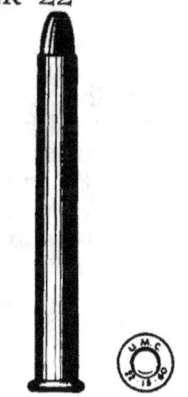

(22-15-60 Stevens)

Total length $2\frac{1}{4}$ in.
2 in. brass straight case
  Dia. at head .266
  Dia. at mouth .247
60 gr. conical, flat nose, lead bullet
15 grs. black powder

In 1896 Stevens introduced this cartridge for their Model 44 rifles. The cartridge itself was designed by Charles H. Herrick of Winchester, Mass. It was originally primed with a No. $1\frac{1}{2}$ primer.

CALIBER 22

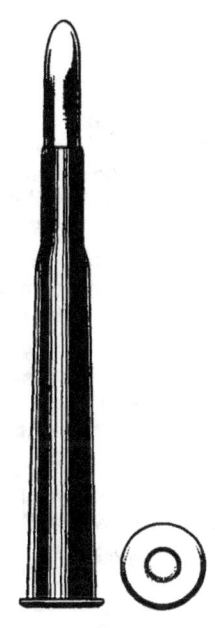

(22 F. A. Experimental)
Total length $3\frac{1}{2}$ in.
$2\frac{3}{4}$ in. tinned, necked case
  Dia. at head .420
  Dia. at mouth .250
120 gr. metal jacket bullet .224 dia.

In December of 1896 Frankford Arsenal experimented with this .22 caliber cartridge . . . with the idea of a smaller service caliber.

CALIBER 220

(220 Swift)
Total length 2²¹⁄₃₂ in.
2³⁄₁₆ in. brass, necked case
Dia. at head .442
Dia. at mouth .255
46 gr. mushroom bullet

Brought out in 1935, the 220 Swift is the highest velocity commercial cartridge ever produced in this country. It is used in the Model 70 Winchester rifle. Its case has an extra strong head to withstand the terrific pressure generated by this cartridge upon firing.

REMINGTON
Remington CLEANBORE DuPont
.220 Swift
Hi-Speed
46 grain mushroom bullet
20 Center Fire Smokeless Cartridges
Made in U. S. A.

CALIBER 230

(230 German C. F.)
Total length 21⁄32 in.
7⁄16 in. brass case, .226 dia.

This cartridge was manufactured by the German firm of Rheinische Westfalische Sprengstoff. It was designed for small hand guns of the bulldog revolver types.

CALIBER 6 mm.

(6 mm. Lee Straight Pull)
Total length 3³⁄₃₂ in.
2¹¹⁄₃₂ in. brass, necked case
Dia. at head .441
Dia. at mouth .276
112 gr. metal jacket, soft point bullet

The cartridge illustrated here is the one used in a Winchester sporting model which was produced from 1897 to around 1902. The 6 mm. Lee was one of the smallest caliber Military cartridges ever used by the Navy. It is said that Winchester claims this to be the first really successful rimless cartridge produced in this country. The Lee Straight Pull U. S. Navy rifle model 1895 was used by the Navy during the Spanish-American War. Three firms are thought to have produced rifles of this bore . . . Winchester, Remington and Blake. Colt also produced a machine gun of 6 mm. caliber.

CALIBER 236

(236 USN)
Total length 3¾16 in.
2⅚₁₆ in. brass, necked case
Dia. at head .441
Dia. at mouth .265
112 gr. metal jacket bullet

According to Saterlee's *A Catalog of Firearms for Collectors*, Blake made a 6 mm. magazine rifle to use a rimmed cartridge. Whether or not any other rifle used this cartridge is not known. The cartridge itself is a desirable collector's item today . . . and their scarcity would indicate that not too many of them were produced.

CALIBER 25

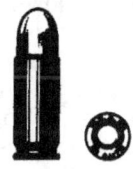

(25 ACP—Automatic Colt Pistol)
Total length ⅞ in.
1⁹⁄₃₂ in. brass case, .277 dia.
50 gr. metal jacket bullet

John M. Browning designed this cartridge in 1906 . . . although Colt did not produce a gun for it until 1908. In Europe this cartridge is known as the 6.35 Browning. A current label reads:

50   WINCHESTER   50
            .25 (6.35 mm.)
AUTOMATIC PISTOL FULL PATCH
SMOKELESS   50 gr
CARTRIDGES   Full Patch
                          Bullets
Adapted to Colt, Browning, Clement
    and other Automatic Pistols
207   12-13
        Mfg. by the
WINCHESTER REPEATING ARMS
          COMPANY

CALIBER 25

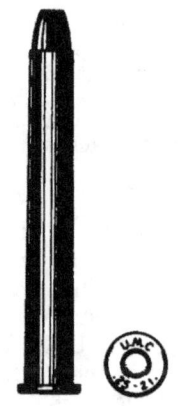

(25-21 Stevens)
Total length 2⁹⁄₃₂ in.
2¹⁄₃₂ in. brass case
Dia. at head .300
Dia. at mouth .283
86 gr. conical, flat nose, lead bullet
21 grs. powder

This cartridge is a modification of the 25-25 Stevens. They derive the name Stevens because of the fact that Stevens Arms Company was the first to make rifles for them. The 25-21 cartridge was used in both the Stevens Ideal No. 44½ and the Stevens Ideal Modern Range No. 47 rifles.

CALIBER 25

(25-25 Stevens)
Total length 2⅝ in.
2⅜ in. brass case
  Dia. at head .299
  Dia. at mouth .281
86 gr. metal jacket, soft point, bullet
25 grs. powder

Originally made up by the Ideal Manufacturing Co. and introduced in Stevens rifles. This cartridge was developed at the suggestion of Capt. W. L. Carpenter, U. S. Infantry. It too, was used in the Models 44½ and 47 Stevens rifles.

CALIBER 25

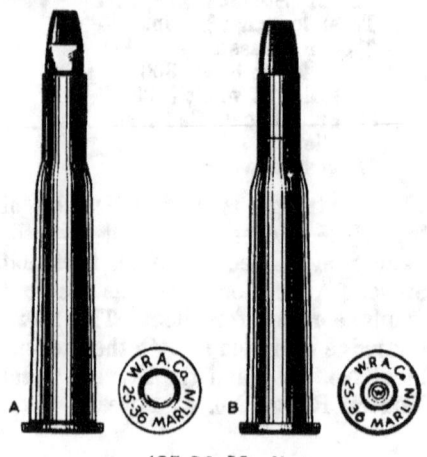

(25-36 Marlin)
Total length 2⁷⁄₁₆ in.
2⅛ in. brass, necked case
  Dia. at head .421
  Dia. at mouth .277
106 gr. cylindrical, flat nose, lead bullet
36 grs. powder

An obsolete cartridge which was used originally in the Marlin repeating rifle, Model 1893. It was developed to compete with the 25-35 Winchester. Although the two cartridges are very similar, they are not interchangeable.
A—Loaded with black powder.
B—Loaded with smokeless powder.

CALIBER 6.5 mm.

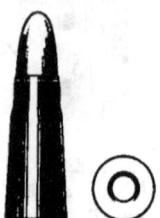

(6.5 Bergmann, rimless, grooveless)
Total length 1⁷⁄₃₂ in.
²⁷⁄₃₂ in. brass, grooveless, rimless case
  Dia. at head .373
  Dia. at mouth .291
Metal jacket bullet

This early automatic cartridge was manufactured by Deutche Metalwaffen, Karlruhe, for the Bergmann auto pistol. The Bergmann, brought out in 1894, was one of the first automatic (auto loading) developed. A 5 mm. size of this same cartridge was one of the smallest center fire cartridges made at that time. The force of its explosion blows this grooveless and rimless case out of the chamber. (See plate 145 *Hand Cannon to Automatic*)

CALIBER 6.5 mm.

(6.5 Bergmann)
Total length 1³⁄₁₆ in.
²⁷⁄₃₂ in. brass, necked case
  Dia. at head .366
  Dia. at mouth .290

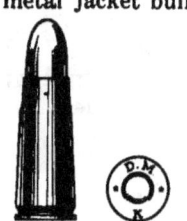

76 gr. metal jacket bullet

This is the common type Bergmann cartridge. It was made in the 5 mm., 6.5 mm., 7.65 mm., and the 8 mm. calibers. A few may be found with lead instead of metal jacket bullets. The Bergmann automatics have not been manufactured for over forty years.

CALIBER 6.5 mm.

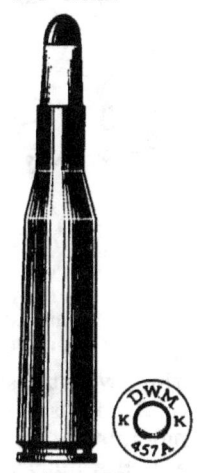

(6.5 x 54 short Mauser)
Total length 2²¹⁄₃₂ in.
2⅛ in. brass, necked case
Dia. at head .467
Dia. at mouth .287
Metal jacket, soft point bullet

Illustrated here is one of the small Mauser cartridges. It should be noted that DWM (Deutsche Waffen und Munitions Fabriken) use numbers instead of the usual caliber designations on headstamp. In this instance the number 457A means 6.5 x 54. Some of the other DWM markings will be found in the Appendix section.

CALIBER 6.5 mm.

(6.5 mm. Italian Carcano)
Total length 3 in.
2¹⁄₁₆ in. brass, necked case
Dia. at head .446
Dia. at mouth .296
162 gr. cupronickel jacket . . . round nose bullet
34 grs. powder

This was the Italian service cartridge of World War II which was used in the Italian Mannlicher-Carcano rifle. It was likewise used in carbines and in an automatic rifle called the Revelli. The Italian Breda machine gun was also chambered to use this load.

CALIBER 7 mm.

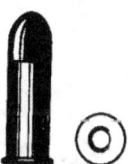

(7 mm. German Bar)
Total length ²⁹⁄₃₂ in.
⅝ in. brass case, .273 dia.
Cylindrical, round nose bullet

Type of cartridge used in the two barrel, four shot, German Bar pistol. It is said that a leading mail order house of this country sold these arms quite extensively from 1900 to 1905.

(See plate 142 *Hand Cannon to Automatic*)

CALIBER 7 mm.

(7 mm. Baby Nambu)
Total length 1 1/16 in.
25/32 in. brass, necked case
    Dia. at head .354
    Dia. at mouth .299
Metal jacket bullet

This is the scarce Baby Nambu cartridge used in the small Baby Nambu automatic. While these arms were used by some of the Japanese officers, they are very rare souvenirs of World War II.

CALIBER 7 mm.

(7X57 mm. Mauser Rifle)
Total length 3 1/16 in.
2 7/32 in. brass, necked case, rimmed
Metal jacket, soft point, bullet

One of the early type Mauser heads, with the raised letters. Made by Deutsche Waffen Und Munitions, Fabriken, Germany. There are two other types of the 7 x 57 . . . one being a flat base, rimmed type . . . the other the 7 mm. rimless.

CALIBER 276

(.276 Pedersen)
Total length 2 27/32 in.
2 in. brass, necked case
    Dia. at head .448
    Dia. at mouth .313
Metal jacket, pointed bullet

This cartridge with the small nickeled primer No. 2 was used in the experimental semi-automatic Pedersen rifle. This rifle used a lubricated case and bullet. The same cartridge with the larger primer was used in the experimental Garand rifle.

A box label reads as follows:

    20 Cartridges, Ball
    Caliber .276
Powder, IMP 25 Lot 1324
                  Cartridge Lot 23
Empty cartridge cases should be cared for and disposed of as prescribed in paragraph 29, A. R. 775-10.
Drawings FB 9892 and PC 50
Manufactured at Frankford Arsenal, 1929

CALIBER 28

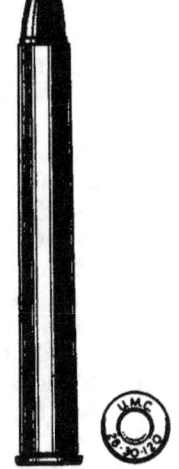

(28-30-120 Stevens)
Total length 2¹³⁄₁₀ in.
2½ in. brass, tapered case
    Dia. at head .357
    Dia. at mouth .309
120 gr. conical, flat nose, lead
    bullet
30 grs. powder

Charles H. Herrick of Winchester, Mass., designed this cartridge, which was first introduced by the Stevens Arms Co. It was used in their No. 44½ and 45 model rifles.

The cartridge was a favorite with Harry M. Pope and other fine barrelmakers. It was not only a very popular target cartridge, but a most accurate one as well.

CALIBER 280

(280 Ross)
Total length 3½ in.
2⁹⁄₁₆ in. brass, necked, semi-rimmed case
    Dia. at head .529
    Dia. at mouth .317
145 gr. steel jacket bullet—
    hollow point
Loaded with nitrocellulose
    powder

This number was used in the Canadian Ross sporting rifle, Model of 1910, and the Ross match rifle. There has been much controversy over the merits of the Ross rifles. The cartridge itself, however, has been quite popular in England. It is said by some to be one of the most interesting high velocity, high power cartridges ever developed.

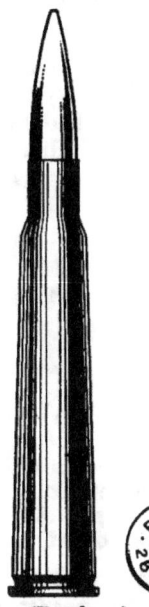

Quoting from a label:

    .280 ROSS SPORTING
    20   .280
  Cartridges   Calibre
  145 grain
  Steel Jacket   Smokeless
    Bullet   Powder
Expanding Bullet with Steel Jacket
  Made and guaranteed by the
  Ross Rifle Co., Quebec, Canada

CALIBER 7.5

(7.5 Swiss Army)
Total length 1⅜ in.

⅞ in. brass, tapered case
Dia. at head .356
Dia. at mouth .319
Metal jacket bullet

Cartridge used in the Model 1882 Swiss Army revolvers. Note the beveled head. The early manufacture of these cartridges had heavy lubricated bullets. The bullets of recent German manufacture were not lubricated.

CALIBER 30

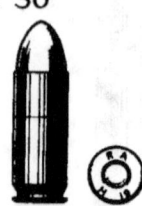

(30 Pedersen Device)
Total length 1 3/32 in.
25/32 in. brass case, .334 dia.
Metal jacket bullet

An unpretentious looking cartridge, yet one of the top secrets of World War I. J. D. Pedersen, an employee of Remington Arms, developed a device, which when attached to the regulation Springfield rifle, would make of it an automatic rifle. This device was in reality an automatic bolt which could replace the regular Springfield bolt. Attached to it was a magazine holding 40 of these small .30 caliber cartridges. A section of the bolt was rifled, thus starting the bullet on its way.

Only government museums have specimens of the device. The rest were ordered destroyed following the war. Even though some 65,000 were manufactured, they did not see service.

The cartridge boxes were labeled to fool any enemy agent as to the actual use of the cartridge . . . and even the workers in the factory did not know in what guns the cartridges were to be used.

A paper box containing enough cartridges for one clip was labeled:

40 CAL. .30 AUTO PISTOL
BALL CARTRIDGES
MODEL OF 1918
THE REMINGTON ARMS
UNION METALLIC CARTRIDGE
COMPANY, INC.
Bridgeport, Conn.

It is stated upon good authority that the French MAS 7.65 long auto pistol is chambered for this cartridge . . . but its use is not advocated.

There was another cartridge made for the Pedersen device which is quite rare. It has a longer bullet and is the one which is identical in appearance to the 7.65 long Auto French MAS. The label from an original box reads . . .

30-18 AUTOMATIC
50 Center Fire Cartridges
Trademark   Remington   90 gr.
Patents   U M C   Metal Cased
                         Bullet
THE REMINGTON ARMS
UNION METALLIC CARTRIDGE
COMPANY, INC.
Bridgeport, Conn.

CALIBER 7.62

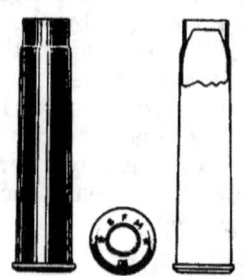

(7.62 Russian Nagant Revolver)
Total length 1 17/32 in.
108 gr. flat nose, metal jacket, bullet seated inside the case
Dia. at head .356
Dia. at mouth .289
12 grs. smokeless powder

As the hammer is cocked on the Nagant revolver, the cylinder is thrust forward to project the cartridge case slightly into the barrel. This makes a gas-tight fit when the cartridge is fired, and is no doubt the reason for its being called the "gas seal cartridge." The revolver was the invention of Nagant, a Belgian. It was adopted by the Russian government in 1895. The cartridge here illustrated is of French manufacture.

CALIBER 7.63

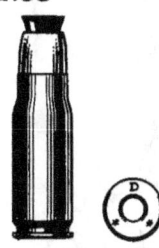

(7.63 German Mauser)
Total length 1 5/16 in.
31/32 in. brass, necked case
  Dia. at head .389
  Dia. at mouth .328
Metal jacket, soft nose, bullet

A very unusual type cartridge in which the lead soft point serves as a wedge to expand the bullet upon impact.

CALIBER 7.65

(7.65 Borchardt Auto)
Total length 1 11/32 in.
1 in. brass, necked case
  Dia. at head .390
  Dia. at mouth .320
Metal jacket bullet

The grand-daddy of the modern automatic (auto-loading) was the Borchardt . . . an invention of a Connecticut Yankee, Hugo Borchardt. Being a "prophet without honor in his own country," he took his invention to Germany. The Germans not only accepted his newy developed arm, but employed him as an engineer. The Borchardt's were first placed on the market in 1893.

So well did Borchardt design the cartridge to go with his automatic, it set the pace for automatic cartridges from that time on. (The 7.65 is identical with the 7.63 in size.)

The Borchardt was redesigned in 1900 and appeared as the well known Luger. (See plate 144, *Hand Cannon to Automatic*)

CALIBER 7.65

(7.65 Hungarian Frommer, short)
Total length 27/32 in.
1/2 in. brass case, .335 dia.
Metal jacket bullet

This small cartridge was used in the first model Hungarian Frommer . . . an arm that is very rarely seen in this country. Developed in 1901, the arm is the predecessor of the present Frommer-Stop.

Among some collectors this cartridge is known as the Roth-Sauer, since it also fits this arm. It is a rather difficult cartridge to obtain, even for a collector.

A similar cartridge was manufactured by Winchester in this country for a very short period. They were head stamped . . . 7.65 Roth-Sauer.

CALIBER 30

(30 M1 Carbine)
Total length 1 21/32 in.
1 9/32 in. brass case
  Dia. at head .353
  Dia. at mouth .331
110 gr. metal jacket bullet
14 gr. powder

A development of Winchester at the

request of the Ordnance Dept. . . . It was used in the new Winchester M1 Carbine of World War II. This arm was used by paratroopers, rangers, engineers and signal corps personnel. A forerunner of the above cartridge, as put out by Winchester was called "Cal. .30 Short Rifle M-1, Self Loading." This "commercial version" was headstamped as follows . . . W.R.A. .30 S. L.

CALIBER 300

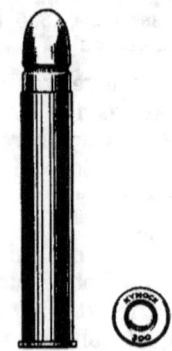

(300 bore long Kynoch)
Total length 2 in.
$1^{17}/_{32}$ in. brass case, .316 dia.
120 gr. copper capped bullet
8½ grs. powder

This cartridge has a hollow point lead bullet with a copper cap and is metal cased ⅞ of its length. It was used in the English Sherwood rifle and is a bit scarce in this country.

CALIBER 30

(30 Krag-rimmed)
Total length 3⅛ in.

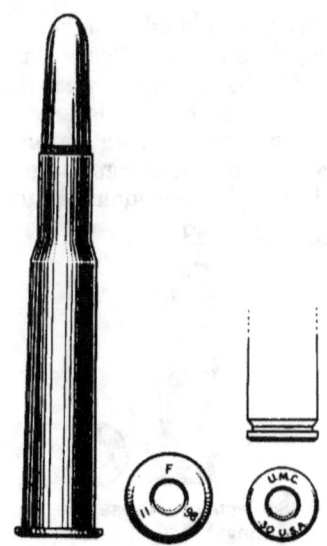

$2^{5}/_{16}$ in. tinned, necked case
Dia. at head .459
Dia. at mouth .333
220 gr. metal jacket bullet
40 grs. powder

This was the Spanish war cartridge used in the U. S. Krag Jorgensen rifle. Originally, they were called "30 U. S. Government," then "30 U. S. Army" and "30 Army" and finally "30 U.S.A." Krags were used in the service from 1894 until around 1903. It saw service as a training rifle during World War I. Many of them, converted to sporters, are in use today by sportsmen.

A similar cartridge, only rimless, was used in one model of the Blake rifle. This was the first sporting rifle in America with a central magazine. The "rimless Krag" was an experimental army cartridge which never was produced to any extent commercially.

CALIBER 7.62 mm.

(7.62 Russian)
Total length 3 in.
$2^{1}/_{16}$ in. brass, necked case
Dia. at head .487
Dia. at mouth .335
148 gr. cupronickel jacket—

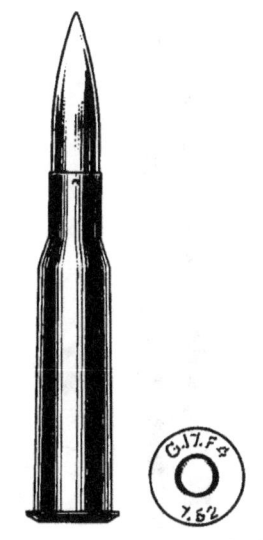

pointed bullet with hollow base
50 grs. powder

The service ammunition used in the Russian MOISIN-NAGANT rifle during World War II. The letters and figures on the beveled head are of the raised type.

CALIBER 30

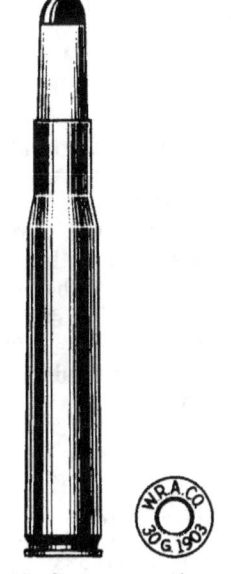

(30-03 Government)

Total length 3¹¹⁄₃₂ in.
2¹⁷⁄₃₂ in. brass, necked case
   Dia. at head .468
   Dia. at mouth .338
220 gr. metal jacket, soft point, bullet

After a great deal of experimenting with rifles employing the Mauser type of action, the government finally settled on the "old favorite" — the Model 1903 Springfield Rifle. Cartridges were produced both in full metal jacket and metal jacket with soft point. These cartridges were only in service for a comparatively short time before being replaced by the popular "30-06."

CALIBER 30

(30-06 Springfield)
Total length 3⁵⁄₁₆ in.
2¹⁵⁄₃₂ in. brass, necked case
   Dia. at head .468
   Dia. at mouth .335
175 gr. gilding metal jacket pointed bullet
50 grs. smokeless powder

Hardly any description is needed of this American cartridge. Used since 1906 ("06"), through two World

Wars, it is the best known of any military cartridge. The Springfield rifle for which this ammunition was developed is still considered one of the finest and most accurate military rifles in the world.

CALIBER 7.7 mm.

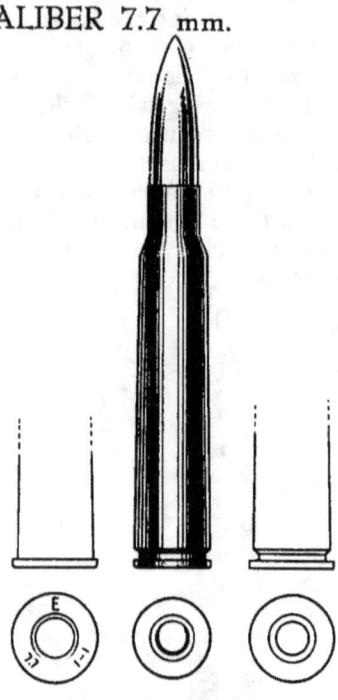

(7.7 mm. Japanese)
Total length 3⅛ in.
2¼ in. brass, necked case
Metal jacket bullet

This cartridge replaced the 6.5 mm. Japanese service caliber which was used in the early part of World War II. It was made in the rimless, rimmed and semi-rimmed types. The rifle uses only the rimless ammunition. The semi-rimmed and rimmed were used in machine guns. The cartridge is very similar to the .303 British service cartridge . . . so far as bullet weight and powder charge are concerned.

CALIBER 303

(303 British)

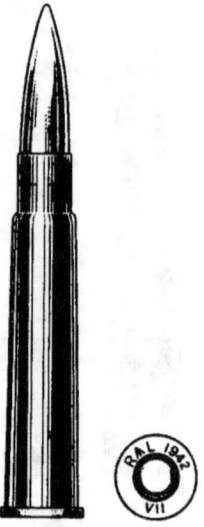

Total length 3¹/₃₂ in.
2⁵/₃₂ in. brass, necked case
Dia. at head .456
Dia. at mouth .338
174 gr. cupronickel jacket bullet
38 grs. cordite powder

The cordite powder used in this British service cartridge is in the form of small sticks. In looking at a cutaway view of the cartridge, the powder looks very similar to a large cable with the strings of powder running lengthwise and slightly twisted. The cartridge here illustrated was manufactured by the Royal Laboratories, Woolwich. England has used the .303 caliber in her military rifles since around 1890.

While the number VII appears after the word Mark, it is but one of a variety of numbers to be encountered . . . Mark V, Mark VI, etc.

The British also made a .303 Rimless . . . experimentally used in the Lewis machine gun.

CALIBER 30

(.30 Newton)
Total length 3⁵/₁₆ in.
2½ in. brass, necked case
Dia. at head .523
Dia. at mouth .336

170 gr. hollow point, copper jacket bullet

Newton cartridges were manufactured by the old U. S. Cartridge Co., Remington and later by Western. They were used in the Newton arms, a development of Charles Newton. Previous to the .300 and .375 H & H Magnum, the .30 caliber was the most powerful American game cartridge. Originally designed for a Fred Adolph of Genoa, N. Y., by Newton, it was for some time referred to as the Adolph Express.

Two other Newton cartridges which may be encountered are the .256 Newton and a .35 Newton.

CALIBER 310

(310 Cattle Killer)
Total length 1 5/16 in.

2 7/32 in. brass case
Dia. at head .354
Dia. at mouth .321
Cylindrical, round nose, lead bullet

These cartridges are used in a short barrel which is held in the hand while the cartridge is detonated by a blow on the firing pin with an ordinary mallet. A label from one of the boxes reads:

200
.310 CATTLE KILLER
CARTRIDGES
SMOKELESS
Manufactured at the
KYNOCH FACTORIES OF IMPERIAL CHEMICAL INDUSTRIES,
LIMITED
London, England

For W. W. Greener Ltd., Birmingham
For Greener Humane Cattle Killer Pistol

CALIBER 7.92

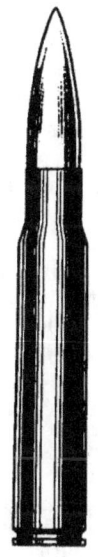

(7.92 German Mauser)
Total length 3 3/16 in.
2 1/4 in. brass, necked case
Dia. at head .471
Dia. at mouth .355
154 gr. metal jacket bullet— hollow base
50 grs. powder

Mauser cartridges adopted for the Mauser rifle in 1898 have undergone little change in outward appearance. After 1905 the original round nose bullet was changed to the present pointed type. The Mauser cartridge is perhaps one of the most widely copied or adapted military cartridges known.

This is the German service cartridge of World War II. A similar cartridge using a solid base Spitzer type bullet was used in World War I.

A popular commercial cartridge, of the above type, was made in the 8 mm. size with a soft point.

CALIBER 8 mm.

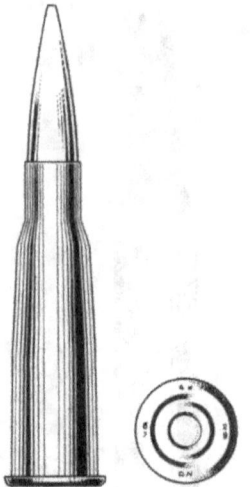

(8 mm. Gaulois)
Total length 9/16 in.
11/32 in. brass case, .313 dia.
Metal jacket bullet

This small cartridge was used in the Gaulois repeating pistol. This was one of the squeezer type palm pistols which were the forerunners of the present day automatics. Squeezer type arms were for the most part made in France, Belgium, and the U. S. The Gaulois was of French manufacture. (See plate 138, *Hand Cannon to Automatic*)

CALIBER 8 mm.

(8 mm. French Lebel)
Total length 2 15/16 in.
1 31/32 in. brass, necked case
    Dia. at head .542
    Dia. at mouth .349
198 gr. solid, bronze alloy bullet—boat tail base
46 grs. powder

The 8 mm. French Lebel was the first small bore military rifle. It was adopted in 1886 in the black powder days. In 1898 a boat tail type of bullet was perfected. The Lebel was also one of the first to use smokeless powder.

This interesting cartridge was used in World Wars I and II as the official service cartridge of France.

CALIBER 8.1 mm.

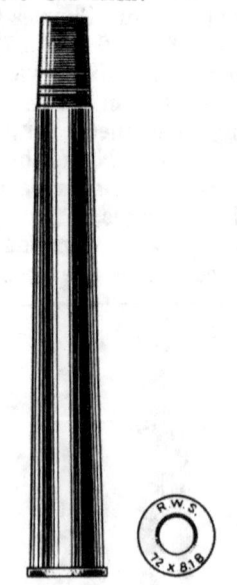

(8.1 x 72 Rimmed)
Total length 3 3/8 in.

2¹³⁄₁₆ in. brass tapered case
Dia. at head .428
Dia. at mouth .343
Flat nose, lead bullet
46 grains powder

An early German sporting cartridge produced by RWS (Rhenische Westfalische Sprengstoff). It will be noted on the headstamp of this particular cartridge that the usual two figures, denoting caliber and case length, are reversed and the length of the case is given first and the caliber second.

CALIBER 32

(.32 Protector)
Total length ⁹⁄₁₆ in.
¹¹⁄₃₂ in. brass case
40. gr. round nose, lead bullet
4 grs. black powder

A small short cartridge used in the Protector Palm pistol manufactured by the Minneapolis Fire Arms Co. These palm pistols were in use during the "gay nineties." They derived their name from the fact that they were designed to fit in the palm of the hand with the barrel protruding between the index and middle fingers. Opposite the barrel was a curved lever which, when pressed in from the squeezing action of the hand, tripped the firing pin and detonated the cartridge. (See plate 140, *Hand Cannon to Automatic*)

CALIBER 32

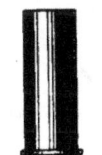 

(32 Merwin & Hulbert)
Total length ⅞ in.
Brass case, .334 dia.
Round, lead ball

6 grs. powder

This is the gallery load for the Merwin & Hulbert 7-shot automatic-ejecting revolver. This arm came out around 1885. A label from an old box of these interesting cartridges gives this information.

50 CARTRIDGES
for
MERWIN & HULBERT CO.
32 cal. 7-shot AUTOMATIC
Round Ball - Six Grains Powder
Manufactured by
AMERICAN METALLIC CARTRIDGE CO.
So. Coventry, Conn.

CALIBER 32

(32-44 S & W Target)
Total length 1¹⁄₃₂ in.
³¹⁄₃₂ in. brass case
Dia. at head .348
Dia. at mouth .348
83 gr. conical, lead bullet
10 grs. black powder

Two types of this cartridge—one containing a conical, lead bullet which was used for target shooting—the other used for gallery shooting—contained a round ball of 50 grains weight. These cartridges were used in a special target revolver brought out by Smith & Wesson as a companion arm to their 38-44. Manufacture of both arms was discontinued in 1910.

CALIBER 32

(32 Extra Long)
Total length 1¹³⁄₁₆ in.
1¼ in. brass case, .317
115 gr. cylindrical, lead bullet
20 grs. powder

The No. 2 Ballard sporting rifle was

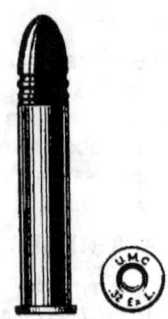

introduced in 1883. While it was chambered for the .32 long rim fire or center fire, it could be chambered on order for the .32 extra long cartridge.

50   .32 Cal. EXTRA LONG
Center Fire
C A R T R I D G E S
for   and other
BALLARD   RIFLES
.32 Ex. Long Ballard
with UMC no. 1½ Primer
Mfg. by
THE UNION METALLIC CARTRIDGE CO.
Bridgeport, Conn., USA

CALIBER 32

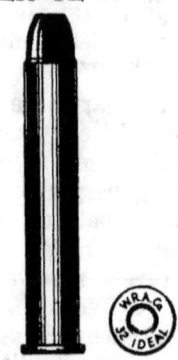

(32 Ideal)
Total length 2 1/32 in.
1 3/4 in. brass, straight case, .345 dia.
150 gr. conical, flat nose, lead bullet
25 grs. powder

Used in both Stevens and Winchester single shot rifles . . . this cartridge was first produced by Union Metallic Cartridge Co. Introduced in 1903, it enjoyed many years of popularity.

CALIBER 32

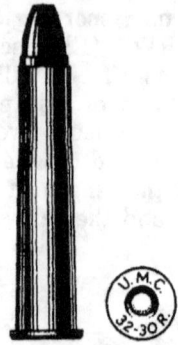

(32-30 Remington)
Total length 2 in.
1 5/8 in. brass, necked case
Dia. at head .377
Dia. at mouth .331
125 gr. conical, flat nose, lead bullet
30 grs. powder

Used in the Remington Hepburn No. 3 sporting rifle. This rifle was built primarily for long range hunting and target purposes. It came out in the early 1880's.

CALIBER 32

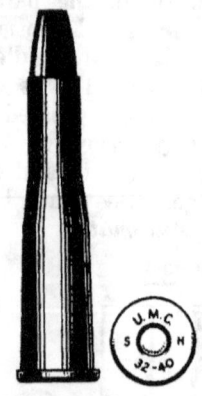

(32-40 Bullard)
Total length 2 1/4 in.
1 27/32 in. brass, necked case
Dia. at head .453
Dia. at mouth .332
150 gr. conical, flat nose, lead

bullet
40 grs. powder

This cartridge is suitable only for the Bullard single shot and repeating lever-action rifles. These rifles were produced along in the 1880's and apparently few were made as they are very difficult to find chambered for this special Bullard cartridge. Winchester and Remington also made a cartridge of this caliber.

CALIBER 32

(32-70 USN)
Total length 3 11/32 in.
2½ in. brass, necked case
   Dia. at head .476
   Dia. at mouth .348
Tinned steel jacket, round nose bullet

One of the many government experimental cartridges. This one was for the navy.

CALIBER 35

(35 S & W Auto)
Total length 3 1/32 in.
2 1/32 in. brass case, .346
76 gr. metal jacket bullet
Manufactured by Remington-UMC

The Smith & Wesson Automatic, manufactured under Clement patents, used this cartridge. The metal jacket on the bullet of this specimen by

Remington UMC has two rectangular slots opposite each other. The lead core fills these slots, thus securely anchoring the metal jacket to the core. These automatics were manufactured from 1913 to 1921. Wording on an original box of cartridges is as follows:

.35 SMITH & WESSON AUTOMATIC,
SMOKELESS
50 CENTER FIRE CARTRIDGES
Metal Point Bullet
THE REMINGTON ARMS UNION
METALLIC CARTRIDGE COMPANY
Incorporated
Ammunition and Firearms
UMC BRIDGEPORT WORKS
Bridgeport, Conn., USA

and on one side:

These Smokeless Cartridges have METAL POINT BULLETS, the lead only coming in contact with the rifling

CALIBER 35

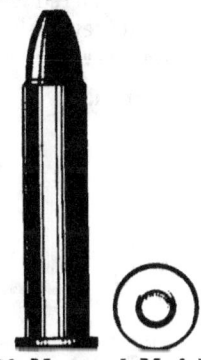

(35-30 Maynard Model 1882)
Total length 2 in.
1 9/16 in. brass, straight case
250 gr. conical, lead bullet
30 grs. black powder

Used in the 1882 series of Maynard's improved hunting and target rifles.

Maynard Hunters Rifle No. 7
Maynard Hunting Rifle No. 9
Maynard Mid-Range Target Rifle No. 10

Since these 1882 Maynards are not head stamped and are nothing like the '65 and '73 types, it is sometimes easy to confuse them with other similar types of this period.

CALIBER 9 mm.

(9 mm. Galand Pistol)
Total length 1 %32 in.
⅞ in. brass case
  Dia. at head .424
  Dia. at mouth .395
Paper patched, lead bullet

C. F. Galand was a gunmaker of Paris. His revolvers are distinguished for their ejecting principle. The trigger guard acts as a lever to throw the barrel and cylinder forward thus permitting the cartridges to be easily removed. It is reported that at one time the Galand revolver was used by the Russian Navy.

CALIBER 9 mm.

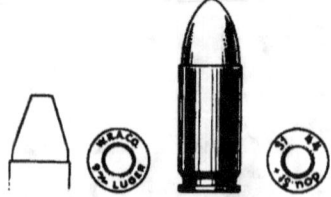

(9 mm. Luger or Parabellum)
Total length 1 %32 in.
¾ in. brass case
  Dia. at head .389
  Dia. at mouth .389
124 gr. conical, metal jacket bullet
5.5 grs. powder

This is the Luger (or Parabellum) military cartridge as manufactured in Germany. Illustrated beside it is one of the Luger commercial cartridges made in the United States. It has a bullet weight of 125 grs.

A similar cartridge but with a dark colored b u l l e t was used in the Schmeisser machine pistol.

CALIBER 9 mm.

(9 mm. Ultra)
Total length 1 in.
$2^{3}/_{32}$ in. brass case, .373 dia.
Conical, metal jacket, flat nose bullet

Made by Gustav Genshow, Durlach-Baden, Germany. This is an experimental cartridge made for the German Luftwaffe. It was designed for a special Walther pistol. It is reported that only 5 pistols and 15,000 rounds of ammunition were made. It is one of the interesting cartridges of World War II.

CALIBER 357

(357 Magnum)
Total length 1½ in.
1 %32 in. nickeled case, .376 dia.
158 gr. wad cutter type, lead bullet

Originally developed for the 357 Magnum revolver, this is the most powerful handgun cartridge manufactured at the present time. It is so powerful that it will penetrate steel plates only dented by other powerful cartridges. It has a muzzle energy of 802 foot pounds and a muzzle velocity of 1,515 feet per second. A label from a box of cartridges by Winchester reads:

```
             WINCHESTER
             Trade Mark
              .357 MAGNUM
    158                    Lead
   Grains                 Bullet
For Smith & Wesson .357 Magnum
             Revolvers
50 Center Fire      Staynless
                  Non-Mercuric
     WINCHESTER REPEATING
             ARMS CO.
   Made in United States of America
```

(See Plate 157, *Hand Cannon to Automatic*)

CALIBER 360

(.360 Mars)
Total length 1¹³⁄₃₂ in.
1 in. nickeled case
160 gr. metal jacket, hollow point bullet

Used in the Mars automatic pistol, an invention of Hugh W. Gabbett-Fairfax, this is the most powerful handgun cartridge ever developed. The gun itself was produced by the Mars Automatic Firearms Syndicate between 1900 and 1902. It was made in three calibers; 8.5 mm., .360 (9 mm.) and .450. The two smaller cartridges used a bottleneck type of case, while the larger one used a straight taper case.

CALIBER 9.3 mm.

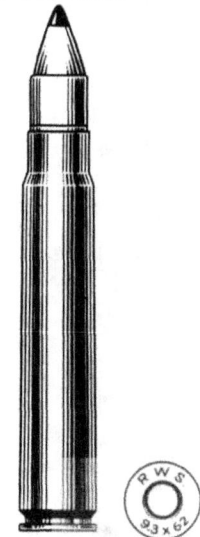

(9.3 x 62 Rimless)
Total length 3⁵⁄₃₂ in.
2¹³⁄₃₂ in. brass, necked case
  Dia. at head .445
  Dia. at mouth .380
258 gr. two piece metal jacket bullet, soft point
60 gr. powder

One of the German big game cartridges. These large cartridges may be found either rimmed or rimless, necked or straight-tapered cases. Hunting cartridges such as these carry two figures, both in millimeters—the diameter of the bullet and the length of the case.

Other sizes in this 9.3 caliber are:
9.3 x 68
9.3 x 72
9.3 x 80
9.3 x 84

CALIBER 9.3 mm.

(9.3 x 80 Rimmed)
Total length 3²¹⁄₃₂ in.
3⁵⁄₃₂ in. brass case
  Dia. at head .428
  Dia. at mouth .385
Flat nose, lead point, metal jacket bullet

Another of the large cartridges produced in Germany for big game hunt-

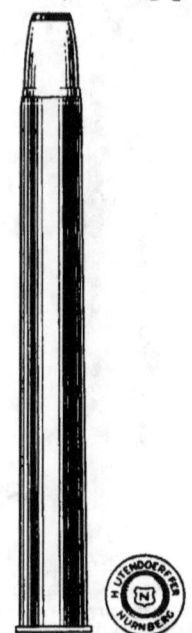

ing. With many early guns and double barreled rifles still in use in Europe prior to World War II, the rimmed cartridge, such as this, was a very popular number.

CALIBER 375

(375 H & H Magnum)
Total length 3½ in.
$2^{27}/_{32}$ in. brass, necked case—belted
    Dia. above belt .511
    Dia. at mouth .397
235 gr. metal, hollow point, expanding bullet

This Holland & Holland is one of the English type hunting cartridges loaded in this country. It is today the most powerful rifle cartridge produced in America. The Winchester Model 70 is the only rifle made commercially to handle this heavy load ... which is used only for big game. A Western label reads:

20 WESTERN
Center Fire Cartridges

Super X
375 H & H Magnum
(Rimless Belted)

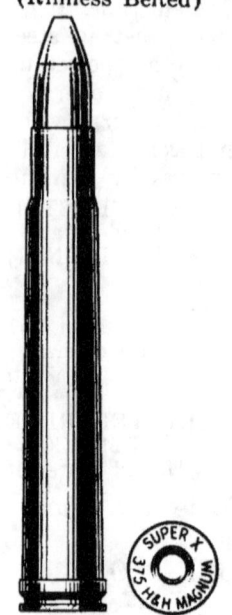

235 Grain   OP. PT. EXP. BULLET
LUBALOY
(Lubricating Alloy)
Smokeless Powder
Non-Corrosive Priming

CALIBER 38

(38 long Colt)
Total length 1⅜ in.
1 in. tinned case .376 dia.
150 gr. round nose, lead bullet
19 grs. black powder

The cartridge here illustrated was made by Frankford Arsenal for use in the Colt .38 Double Action revolver, Models of 1892-94-96. This arm was the first of the side swing cylinder Colts to be adopted by the army.

CALIBER 38

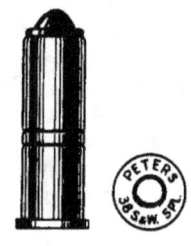

(38 Smith & Wesson Special)
Total length 1 9/32 in.
1 1/8 in. brass case with 2 cannelures, .376 dia.
Wadcutter, lead bullet

This cartridge illustrates the wadcutter bullet of the .38 Special. There are other types of bullets used in the popular 38 Special series. This caliber is a very popular gun with police departments and law enforcement agencies generally.

CALIBER 38

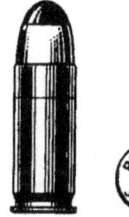

(38 Automatic Colt)
Total length 1 1/4 in.
7/8 in. brass case, .383 dia.
130 gr. metal jacket, soft point bullet

The .38 caliber Automatic Colt Pistol cartridge was the first automatic cartridge developed in this country. It was offered to both the army and navy, but it lost to the heavier .45 which was finally selected. The cartridge illustrated here is the sporting type . . . not the full metal jacket type experimented with by the services. A label from a Remington UMC box reads:

.38 AUTOMATIC COLT SMOKELESS
50 Central Fire Cartridges

REM   38   Soft Point
UMC   Automatic Colt   Bullet
Soft Point
Remington Arms—Union Metallic Cartridge Co.
Ammunition and Firearms
Union Metallic Cartridge Works
Bridgeport, Conn., U. S. A.
Specially adapted for .38 Colt Automatic for Sporting Purposes only

These Smokeless Powder Cartridges are expressly adapted to the .38 Automatic Colt. They being made according to our special directions, we strongly recommend them for use in these arms.
Colt Patent Fire Arms Co.

The .380 Auto is another type of the automatics. (See Plate 148, *Hand Cannon to Automatic*)

CALIBER 38

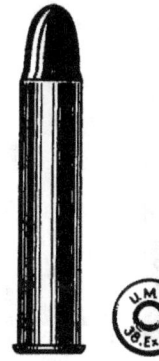

(38 Extra Long)
Total length 2 1/16 in.
1 5/8 in. brass straight case, .379 dia.
146 gr. conical, round nose, lead bullet
31 grs. powder

A cartridge for the early single shot rifles . . . Stevens, Ballards, etc. . . . This description is given by a label:

50   .38 CAL. EX. LONG
CENTRAL FIRE CARTRIDGES
38 EX LONG

Trade
UMC   31 grs. powder
Mark   146 grs. bullet

Manufactured by
THE UNION METALLIC CARTRIDGE CO.

Bridgeport, Conn. USA
.38 EX. LONG BALLARD

CALIBER 38

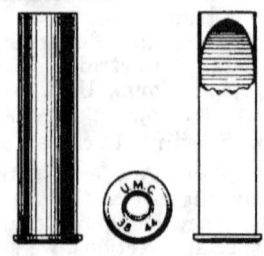

(38-44 Smith & Wesson Gallery)
Total length 1 7/16 in.
Brass case, .383 dia.
146 gr. conical, lead bullet
20 grs. powder

A target cartridge used in the 38-44 Smith & Wesson, heavy duty police and outdoorsman models.

CALIBER 38

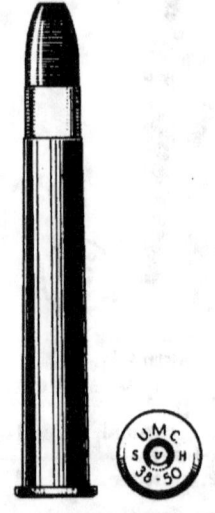

(38-50 Remington-Hepburn)
Total length 3 3/16 in.
2 1/4 in. brass, tapered case
    Dia. at head .453
    Dia. at mouth .392
255 gr. paper patched, lead bullet
50 grs. powder

This special cartridge is oftentimes referred to as 38-2¼ . . . meaning .38 caliber with a 2¼ in. case. The S H on the head stands for solid head to distinguish it from the earlier types of "made up" heads. Solid heads today generally mean those which have a semi-balloon primer pocket projecting up into the powder chamber. The Remington-Hepburn rifles came out around 1880 and were chambered in several calibers. They were the development of Lewis Lobell Hepburn who later became affiliated with Remington Arms.

CALIBER 38

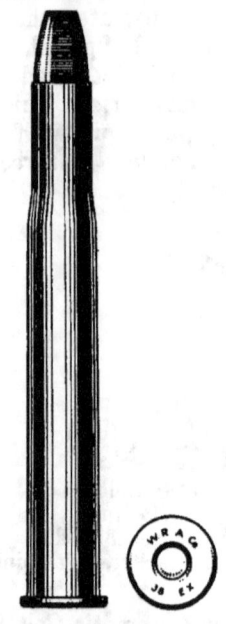

(38-90 Winchester Express)
or 38-3¼
Total length 3 11/16 in.
3¼ in. brass, necked case
    Dia. at head .477
    Dia. at mouth .395
217 grain, flat nose, lead bullet
90 grs. powder

One of the large numbers used in the heavy single shot Winchester high wall rifles. These rifles came out around 1886. The cartridge itself does not seem to have been too numerous after 1900 . . . at least they were not listed in the available calibers after that date.

CALIBER 9.8 mm.

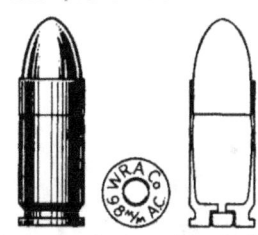

(9.8 mm. Automatic Colt)
Total length 1 19/32 in.
29/32 in. brass case .403 dia.
130 gr. metal jacket bullet

In 1912 Winchester made up some of these cartridges for a few experimental Colts (Model 1911AC) which were being submitted to the Roumanian Government for tests as an army pistol. The Roumanians selected the British Dimancea instead. Today these cartridges are classed as desirable collector's items.

```
     50         WINCHESTER        50
                 Trade Mark
        9.8 mm. Automatic Colt Cartridges
       Adapted to      Smokeless Powder
         Colts              130 grain
     Automatic Pistol Metal Patched Bullets
      Trade Mark Reg. in U. S. Pat. Office
                     and
            Throughout the World
                  Mfg By the
         WINCHESTER REPEATING ARMS
                  COMPANY
```

CALIBER 40

(40-40 Maynard Model 1882)
Total length 2 5/16 in.
1 3/4 in. brass case
  Dia. at head .456
  Dia. at mouth .450
270 gr. conical, flat nose, bullet
40 grs. powder

Maynard rifles could be purchased in cases with one or more interchangeable barrels. The cartridge here illustrated came from such a set. With the .40 caliber rifle barrel, a .64 caliber shot barrel was provided in this cased set.

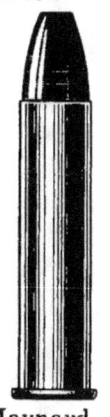

The Maynard Mid-Range Target Rifle No. 10, Model 1882, was supplied in this 40-40 caliber.

CALIBER 40

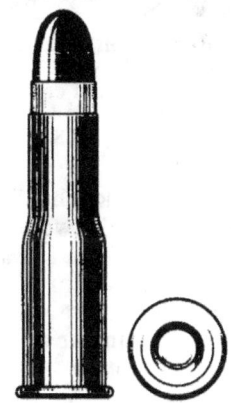

(40-50 Sharps)
Total length 2 5/16 in.
1 11/16 in. brass, necked case
  Dia. at head .505
  Dia. at mouth .429
265 gr. paper patched, lead bullet
50 grs. black powder

Not only did Sharps manufacture these cartridges, but they were also produced by Winchester and UMC. This number was used in the Sharps B. L. Sporting Rifle and the Sharps B. L. Hunters Rifle.

The 40-50 cartridge was also made in a straight case ... which was used in the 1878 Sharps Borchardt Hunters Rifle.

Sharps also loaded this .40 caliber cartridge with 60 grs. of powder—it was called 40-60 Sharps.

CALIBER 40

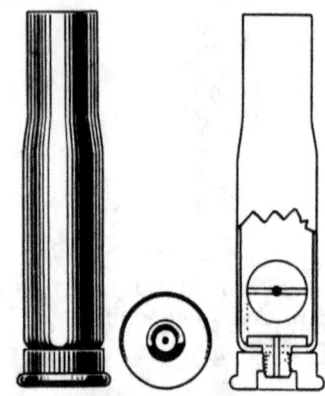

(40-70 Remington Reloading Shells)
2¼ in. brass, necked case with two-piece steel head
Dia. at head .503
Dia. at mouth .435

The cross-section sketch illustrates the construction of this two-piece case. As will be noted, the head is attached to the case by a screw, through which runs the flame hole. The point of the screw serves as an anvil for the primer. A label from a box of these shells reads thus:

25
BRASS SHELLS
for
STEEL HEAD
RELOADING SHELLS
Adapted to
Remington Sporting Rifle
Cal. .40;   70 grs.

CALIBER 40

(40-70 What Cheer)
Total length 2¹³⁄₁₆ in.
1¾ in. brass, necked case with Berdan Primer
Dia. at head .581
Dia. at mouth .429

380 gr. cylindrical, paper patch, lead bullet
70 grs. black powder

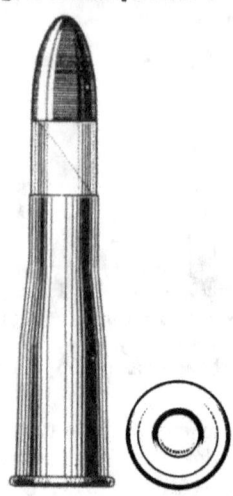

This odd shape cartridge was used in the Peabody-Martini "What Cheer" rifle. The Peabody rifle was invented by Henry O. Peabody in 1862. Later the action was improved by a Swiss inventor by the name of Martini ... hence the name Peabody-Martini.

According to an old issue of *Forest & Stream* (Oct. 1875), a rifle range was opened only seven miles from Providence, R. I., which was named "What Cheer." It is said that Roger Williams, whom the Puritans had banished to the wilderness which later became Rhode Island, and his Indian friends used the words "What Cheer" as a sort of pass word or greeting. It was the Providence Tool Co., makers of the Peabody-Martini long range rifles who instigated the honoring of the "What Cheer" range and so named the rifles and cartridges by the same name. Three calibers were named for the range, 40-70, 40-90, and 44-95. The range opened Oct. 25, 1875.

CALIBER 40

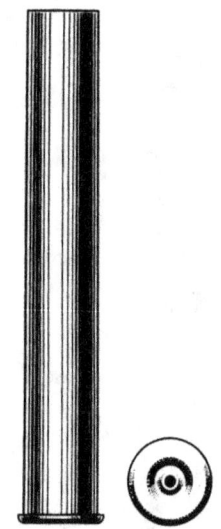

(40-3 1/16 in. Ideal Everlasting Case)
3 1/16 in. brass, tapered case
    Dia. at head .478
    Dia. at mouth .448

These cases were used for individual hand loading. As originally loaded, this case used 90 grs. powder and 370 gr. paper patched bullet. The Pacific Ballard No. 5 was chambered for this not too familiar case. Other rifles have also been found chambered for this shell.

CALIBER 40

(40-85 Ballard)
Total length 3 13/16 in.
2 15/16 in. brass, tapered case
    Dia. at head .477
    Dia. at mouth .424
370 gr. paper patched, lead bullet
85 grs. black powder

This cartridge and the 40-90 Ballard are in reality one and the same. Some will be marked one way and some the other. They were used in the long range hunting Ballard Pacific rifle No. 5 . . . introduced in 1882.

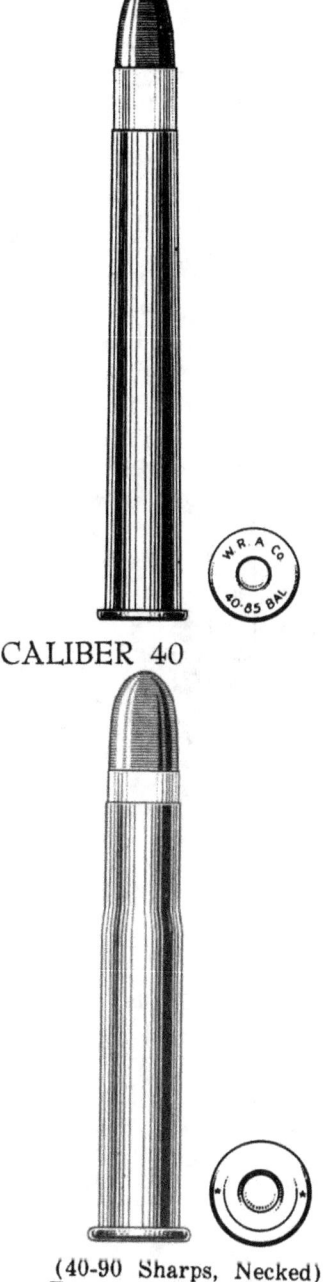

CALIBER 40

(40-90 Sharps, Necked)
Total length 3 7/16 in.
2 5/8 in. brass, necked case
370 gr. paper patched, lead bullet

90 grs. black powder

Used in the early side-hammer Sharps rifles. This caliber was also made in a straight case. These cartridges were introduced about 1876 and were discontinued some four years later.

CALIBER 40

370 gr. paper patched, lead bullet
90 grs. black powder

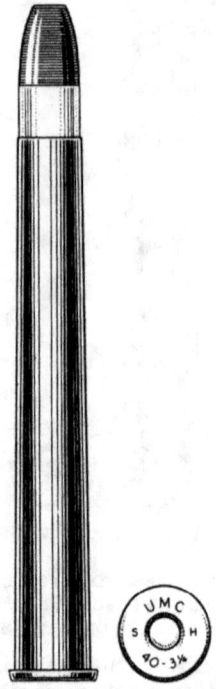

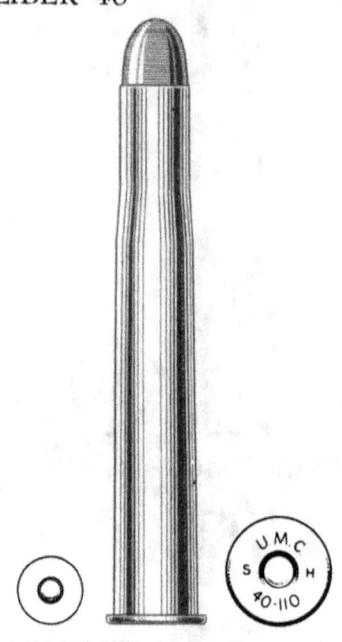

(.40-110 Winchester Express)
Total length 3⅝ in.
3¼ in. brass, necked case
  Dia. at head .543
  Dia. at mouth .428
260 grain express bullet with copper tube
110 grs. black powder

Used in the Winchester single shot rifles of 1885 and after, this cartridge was very effective at short range. A cross-section of the copper tubed Express bullet is illustrated in the Appendix section of this book.

CALIBER 40

(40-3¼ Sharps)
Total length 4⅟₁₆ in.
3¼ in. brass case
  Dia. at head .477
  Dia. at mouth .425

Comes now the big cartridges which saw service in the buffalo days. Although the Sharps Company did not make a cartridge of this case length, other firms such as UMC did produce them for Sharps rifles.

CALIBER 401

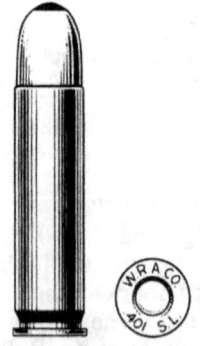

(.401 Winchester SL)
Total length 2 in.
1½ in. brass case

Dia. at head .429
Dia. at mouth .429
200 gr. metal jacket, soft nose,
  copper point bullet
Smokeless powder

This self-loading cartridge was used in the Winchester Model 10 Rifle—a rifle for American large game hunting. The magazine held four cartridges, and with one in the chamber, the hunter had a five-shot repeater. Cartridges were supplied in either 200 or 250 grain, full patch or soft point.

CALIBER 404

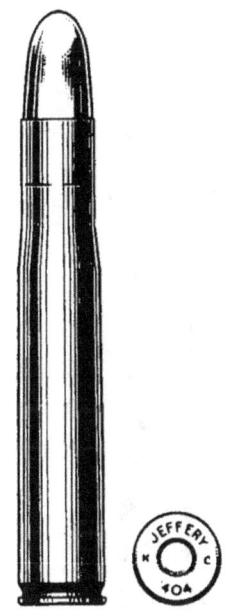

(404 Jeffery Rimless Express)
Total length 3$^{17}\!/_{32}$ in.
2$^{7}\!/_{8}$ in. brass, necked case
300-400 gr. metal jacket bullet

One of the popular English Kynoch cartridges . . . supplied with either a solid metal-covered or copper-pointed bullet . . . loaded with cordite powder as follows:
   70 grs. cordite powder and a 300 gr. bullet . . . metal jacket or copper point
   60 grs. cordite powder and a 400 gr. bullet . . . metal jacket, soft point, split jacket

CALIBER 405

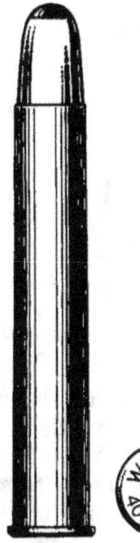

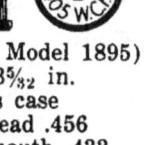

(.405 WCF - Model 1895)
Total length 3$^{5}\!/_{32}$ in.
2$^{9}\!/_{16}$ in. brass case
   Dia. at head .456
   Dia. at mouth .433
300 gr. metal patched bullet
57$^{1}\!/_{2}$ grs. smokeless powder

The lever action repeating rifle Model 1895 was chambered for this cartridge. This rifle was the first to successfully employ the box type magazine to a lever action repeater.

.405 Caliber Metal Patch
          Model 1895
   20   300 Gr.
Cartridges Metal Patched
            Bullet
         Smokeless Powder
Trade Mark Registered in U. S. Patent Office and throughout the world
For Winchester Repeating Rifle, Model 1895
Manufactured by the Winchester Repeating Arms Co., New Haven, Conn.

CALIBER 10.4

(10.4 Swiss)
Total length 2$^{3}\!/_{8}$ in.
1$^{5}\!/_{8}$ in. brass case

Dia. at head .537
Dia. at mouth .443
Paper patched, lead bullet

This is the same as the well known .41 Swiss . . . only this is center fire and of German manufacture, with raised letters on the head stamp. The original box is deep blue in color and has a white label upon which is printed:

20 Stuck Geladene Metallpatronen K 42/10, 4d. Geschoss 42. 3 Cr. & B6.

CALIBER 41

(41 Short)
Total length 1 1/16 in.
5/8 in. brass case .406 dia.
163 gr. round nose, lead bullet
15 grs. powder

This cartridge is now obsolete since no arm chambered for it is being manufactured. They were manufactured up to World II for the many old time guns then still in use.

CALIBER 41

(41 Long Double Action)
Total length 1 13/32 in.
15/16 in. brass case, .407 dia.
200 gr. round nose, lead bullet

21 grs. powder

Designed for the Colt Lightning Model 1877. This was the first double action to be placed on the market by Colt. They were built with and without an ejector rod.

Billy the Kid is said to have died with a .41 Colt Lightning in his hand. An old catalog listed these .41 revolvers as the .41 Colt Thunder . . . while the .38 calibers were called Lightnings.

CALIBER 416

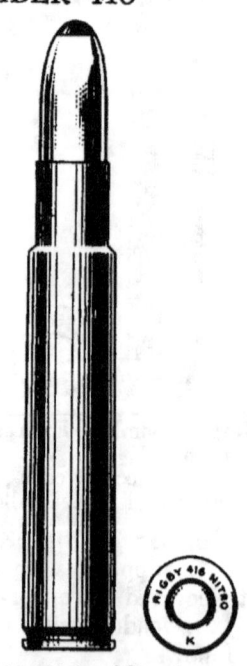

(416 Rigby Nitro)
Total length 3 23/32 in.
2 7/8 in. brass, necked case
410 gr. metal case, soft nose bullet

71 grs. cordite powder

One of the English big game cartridges which is advertised currently in this country.

CALIBER 10.75

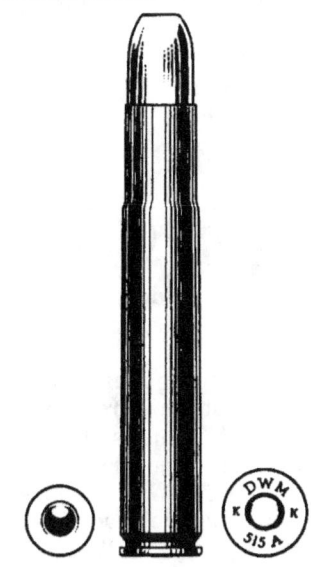

(10.75 x 68 - German)
Total length 3 3/16 in.
2 21/32 in. brass necked case
347 gr. hollow point, metal jacket bullet
65 grs. powder

One of the current German high powered cartridges. A brief explanation of the German system at designating cartridges is interposed here.

The first figure (10.75) refers to the caliber of the bullet. The second figure (68) is the length of the empty brass case. Should the cartridge have a rim, the letter "R" is added to the second figure.

CALIBER 11 mm.

(11 mm. German Revolver)
Total length 1 7/16 in.
3 1/32 in. brass case
  Dia. at head .450
  Dia. at mouth .450
Conical, lead bullet

One of the large caliber revolver cartridges with the early type Mauser head. The letters of the headstamp are of the raised or relief type. This one was made in Germany by G. EGESTORFF—LINDEN.

CALIBER 11 mm.

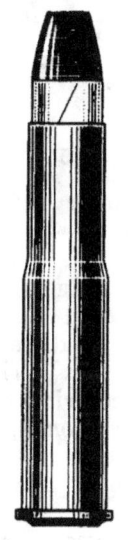

(11 mm. Mauser)
Total length 3 in.
2 11/32 in. brass necked case
  Dia. at head .515
  Dia. at mouth .438
Paper patched, lead bullet
Black powder

Developed along in the late 70's or early 80's and used in a bolt-action, single shot Mauser rifle, this cartridge was used in Germany at the time the 45-70 cartridge was popular in this country.

CALIBER 11 mm.

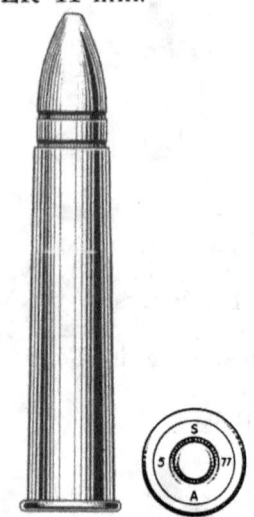

(11 mm. Spanish)
Total length 3 in.
2 3/16 in. brass case
Dia. at head .525
Dia. at mouth .470
395 gr. reformado type of bullet (full metal jacket)
77 grs. powder

These Spanish cartridges were referred to as "Poisoned Bullets" by the United States soldiers in the Spanish-American War. This was due to the brass covering over the bullet corroding and becoming covered with verdigris.

Many types of bullets and various headstamps will be encountered among these cartridges. Remington manufactured a rifle chambered for this number and many of them were sold to the Spanish preceding our war with Spain.

CALIBER 44

(44 Smith & Wesson, American)
Total length 1 13/32 in.
2 7/32 in. brass case, .436 dia.
205 gr. conical, round nose, lead bullet
25 grs. black powder

In 1871 one thousand of the .44 Smith & Wesson revolvers were delivered to the army. Originally named the S & W Army Revolver, it has become better known simply as the 44 S & W American. There were two models with but slight differences.

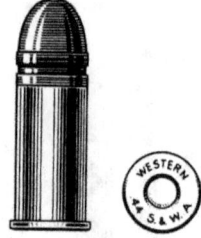

The early cartridges for this arm had a 218 gr. bullet and 28 grs. black powder.

The revolver using these cartridges is one of the historical frontier arms of the old West. This is the arm Buffalo Bill Cody was carrying when he was a guide for Grand Duke Alexis of Russia. The Duke admired it and as a result Smith & Wesson landed a large order for . . . the Smith & Wesson Russian.

A. C. Gould, writing in *Modern American Pistols and Revolvers*, 1888, said:

"The old American model Smith & Wesson revolver was a great favorite with those who knew what weapon to select for reliable work. Many are in use today, and highly valued as very accurate weapons." (See plate 161, *Hand Cannon to Automatic*)

CALIBER 44

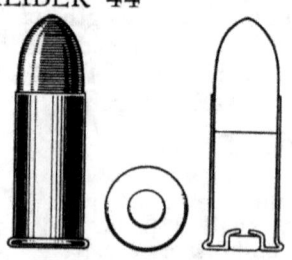

(44 Smith & Wesson Russian)
Total length 1 13/32 in.

³¹⁄₃₂ in. brass case, .453 dia.
246 gr. conical, lead bullet
23 grs. black powder

This cartridge, considered one of the best of its day, was one of the few revolver cartridges to possess both target accuracy and shocking power. It was used in the Model No. 3 single action Russian manufactured by Smith & Wesson. As will be noted from the cross-section view the case of this cartridge was of the early folded, rather than solid rim variety. The revolver was manufactured from 1870 to 1875.

CALIBER 44

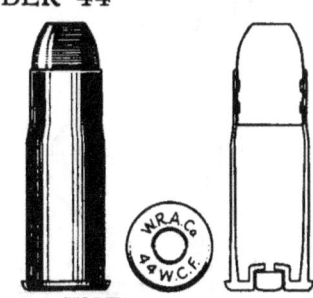

(44-40 WCF)
Total length 1¹⁹⁄₃₂ in.
1⁵⁄₁₆ in. brass case
  Dia. at head .465
  Dia. at mouth .436
200 gr. flat nose, lead bullet
40 grs. black powder

Teamed with the Model 1873 Winchester rifle and the Colt single action revolver, (Frontier Six Shooter), this cartridge not only saw history in the making, but ... helped to make history.

In all probability, during its heyday, it accounted for more game than any other single cartridge. It was with the Winchester when the hunt was on ... it was a companion of the Colt six-gun as marshal, judge and executioner. It permitted the frontiersman to carry only one type of ammunition for both his rifle and revolver ... quite a feature in those days.

The list of those who depended upon this 44-40, or 44WCF, would read like a "Who's Who" of the characters of western history.

From an old box label of these famous cartridges of the Old West, we read:

50　　　　CARTRIDGES　　.44 cal.
for
WINCHESTER RIFLE MODEL 1873
Central　　　　　　　　　　Solid
  Fire　　　　　　　　　　　Head
　　　　Mfg. by the
WINCHESTER REPEATING ARMS
CO.
New Haven, Conn. USA

CALIBER 44

(44 Evans - Old Model)
Total length 1⁷⁄₁₆ in.
⁶³⁄₆₄ in. brass case
  Dia. at head .439
  Dia. at mouth .439
215 gr. conical, flat nose, lead bullet
28 grs. black powder

Thirty-four of these cartridges could be held in the magazine of the old model Evans. This rifle had the greatest magazine capacity of any repeating rifle of American manufacture. It was introduced around 1875 and was discontinued sometime in the 80's.

CALIBER 44

(44 Evans Long - New Model)
Total length 2 in.
1⁹⁄₁₆ in. brass case
  Dia. at head, .449
  Dia. at mouth, .434
280 gr. conical, flat nose, lead bullet
42 grains black powder

This cartridge brings up the point

that different firms loaded the same cartridge differently — for instance note the difference of loading this particular cartridge by four separate companies:

Winchester	44-42-280
UMC	44-43-276
American Metallic Co.	44-43-275
U. S. Cartridge Co.	44-40-300

These cartridges were used in the model 1877 Evans sporting rifle. Twenty-six cartridges could be carried in the magazine located in the stock.

Two types of this item are illustrated.

CALIBER 44

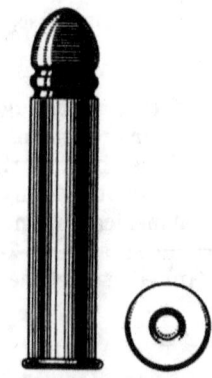

(44 Extra Long)
for Wesson Rifle
Total length 2 3/16 in.
1 5/8 in. brass case with No. 1 Wesson primer, Hobb's Pat.
Dia. at head .440
Dia. at mouth .440
250 gr. round nose, lead bullet
48 grs. black powder

Used in the Frank Wesson long range Creedmore rifle No. 1, which was advertised in the March 3rd, 1877, issue of *Rod and Gun.*

CALIBER 44

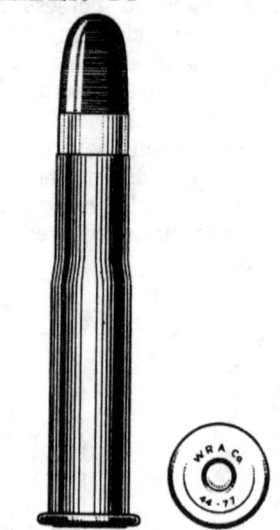

(44-77 Sharps)
Total length 3 3/32 in.
2 1/4 in. brass, necked case
470 gr. paper patched bullet
77 grs. black powder

Used in the Sharps B. L. sporting rifle. A great number of these cartridges must have been made because they are still to be found on the lists of cartridge dealers.

CALIBER 44

(44-95 What Cheer)
Total length 3 3/8 in.
2 5/16 in. brass, necked case
with Berdan primer
Dia. at head .682
Dia. at mouth .467
550 gr. cylindrical, paper
patched, lead bullet
95 grs. black powder

Largest size of the special "What Cheer" cartridges used in the Peabody-Martini "What Cheer" Creedmoor Rifles. On the box label they were referred to as "Special .44 Cal. 'What Cheer' cartridges" with a .95

written in ink after the word "Cal."

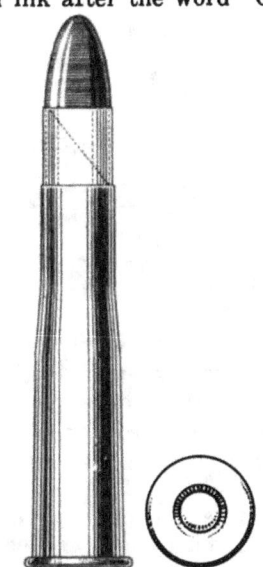

They were manufactured by the Union Metallic Cartridge Company, Bridgeport, Conn., under Berdan's Patents, Mch. 20, '66, and Sept. 29, '68, and S. W. Wood's Patents, Apr. 1, '62 and Apr. 2, '72.

CALIBER 44

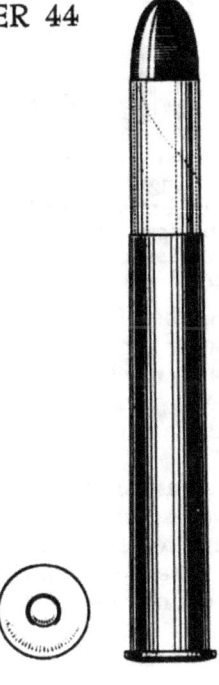

(44-100-550 Remington Creedmoor)
Total length 4 1/16 in.
2 19/32 in. brass case with a lightly curved head
    Dia. at head .503
    Dia. at mouth .469
550 gr. paper patched, lead bullet
100 grs. powder

Used in the Remington-Hepburn No. 3 Long Range Military Creedmoor rifle . . . a gun which was first introduced in 1880. No other rifle is believed to have been chambered for this cartridge, which is becoming increasingly scarce today.

CALIBER 45

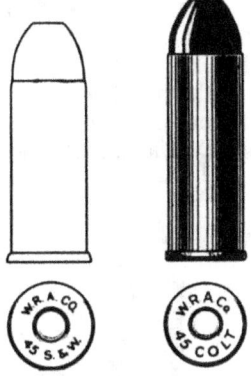

(45 Colt)
Total length 1 19/32 in.
1 1/4 in. brass case, .478 dia.
255 gr. conical, lead bullet
35-40 grs. powder

(45 S & W)
Total length 1 7/16 in.
1 3/32 in. brass case, .474 dia.
250 gr. conical, lead bullet
30 grs. powder

Two contemporary cartridges of Frontier days. The .45 Colt used in the Colt single action army and the .45 S & W in the .45 S & W, Schofield Patent. Both arms made history in the old days and old-timers are still arguing the respective merits of each.

The cartridges are pictured together for comparison.

CALIBER 11.35 mm.

(11.35 mm. Shouboe)
Total length 1⅛ in.
１¹⁄₁₆ in. brass case
   Dia. at head .471
   Dia. at mouth .471
Aluminum covered wood bullet

The unique Shouboe—a Danish experimental cartridge made by DWM of Germany. It is a very light cartridge for its size due to the fact that its bullet is of wood with but a thin aluminum jacket. The Shouboe gun was a Danish military automatic.

CALIBER 45

(45 A.C.P.)
Total length 1¼ in.
²⁹⁄₃₂ in. brass case, .470 dia.
230 gr. metal jacket bullet
5 grs. powder

In 1911 the army adopted the .45 caliber Colt automatic pistol as the official side arm of the service. It has since been known as the Model 1911. The cartridge illustrated here was the type used in this arm . . . and also in the Model 1917 .45 caliber revolver during World War I. It is today the official hand gun cartridge of the United States.

CALIBER 45

(45 Auto Rim)
Total length 1¼ in.
⅞ in. brass case, rimmed, .470 dia.
230 gr. flat nose, lead bullet

This cartridge was developed to use in the 1917 service revolvers which were chambered for the 45 A.C.P. cartridge, which was rimless and required a clip. This cartridge was to eliminate the clip for civilian use. They are not a military load and were made both with a lead bullet and a metal jacket bullet. The amount and type of powder varies with the different manufacturers.

CALIBER 450

(.450 Boxer Revolver)
Total length 1⅛ in.
¹¹⁄₁₆ in. brass case with Boxer type head .475 dia.
225 gr. round nose, lead bullet
13 grs. black powder

In 1868 the British Government issued Boxer type cartridges for the .455 Adams revolver, which was the official side arm. These cartridges were manufactured by Eley Bros. of London.

Since the case of the revolver cartridge was so short, the piece corresponding to the base cup in the regular Boxer cartridge for rifles was merely drawn out to form the entire case. The head, which contained the primer and primer pocket, was riveted on in the same manner as the larger .557. (A glance at the cross-section of the .577 Snider in the Early Center Fires will explain the manner of attaching.) These cartridges will be found with either brass or iron heads.

CALIBER 45

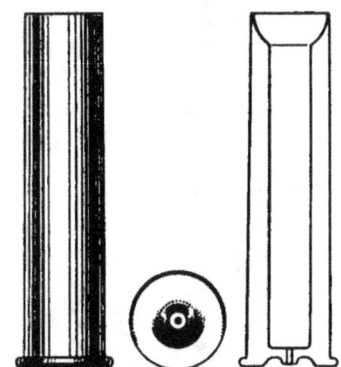

(45 Sharps Steel Cartridge Cases)
Total length $2\frac{1}{10}$ in.
Steel case... Dia. at head .505
Dia. at mouth .480
Round lead ball
12 grs. powder

These solid steel shells, manufactured by the Sharps Rifle Company, were used in gallery and short range shooting. Two sizes are known to have been made ... the size illustrated, and a .40 caliber, 2½ inches in length. Both used a Berdan No. 1 primer. The 1879 Sharps catalog also has this note concerning them. "Their use in military rifles accustoms the soldier to his arm and will perfect him in marksmanship as rapidly as practice with regular military cartridges."

CALIBER 45

(45-60 Straight Target)
Total length $2\frac{7}{32}$ in.
$1\frac{7}{8}$ in. brass, slight necked case
Dia. at head .505
Dia. at mouth .473
300 gr. flat nose, lead bullet
60 grs. black powder

This cartridge was used in the Model 1876 Winchester ... a rifle which

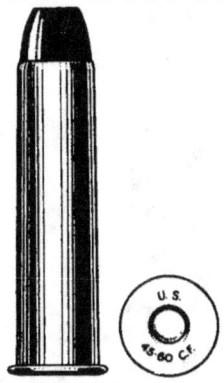

was brought out at the request of sportsmen who desired a repeating rifle of larger caliber than the old 44-40 of the 1873 Model.

CALIBER 45

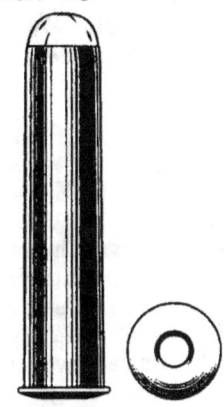

(45-70 Govt.—Phoenix)
Total length $2\frac{9}{32}$ in.
$2\frac{1}{8}$ in. brass case
Dia. at head .502
Dia. at mouth .468
3 balls weighing 108 gr. enclosed in paper container
54 grs. black powder

Made by the Phoenix Cartridge Company. This cartridge is a military guard load. It was adaptable to any of the various rifles chambered for the 45-70 Govt.

CALIBER 45

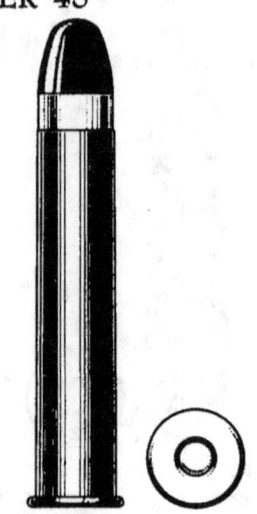

(45-70 Van Choate)
Total length 2²⁹⁄₃₂ in.
2¼ in. brass case
    Dia. at head .506
    Dia. at mouth .473
    420 gr. paper patched, lead bullet
    70 grs. black powder

A quite scarce cartridge today is this 45-70 Van Choate. It was used in the Van Choate B. L. Military Rifle, a patent of S. F. Van Choate. This bolt action rifle was manufactured by the Brown Manufacturing Co. of Newburyport, Mass. Used in the experimental tests of 1872, it is probable that this rifle never got more than past the model stage. This cartridge was still listed in the 1912 Remington-UMC Catalog . . . primed cases and bullets.

CALIBER 45

(45-70-500 Government)
Total length 2²⁵⁄₃₂ in.
2⁵⁄₃₂ in. tinned case
    Dia. at head .503
    Dia. at mouth .478
    500 gr. round nose, lead bullet
    70 grs. powder

Produced in the early 1870's this cartridge was for a good many years the official service long arm ammunition. Several commercial rifles were also chambered for it. It used either a 500- or a 405-grain bullet and was originally loaded with black powder. The cartridge is still obtainable even though no rifle has been made to handle it for quite some time.

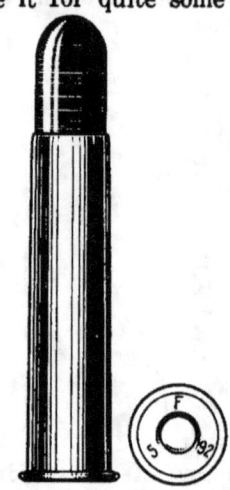

This cartridge with the 405 gr. bullet was in use by the 7th Cavalry under Custer when they were wiped out at the Battle of the. Little Big Horn on June 25, 1876.

CALIBER 45

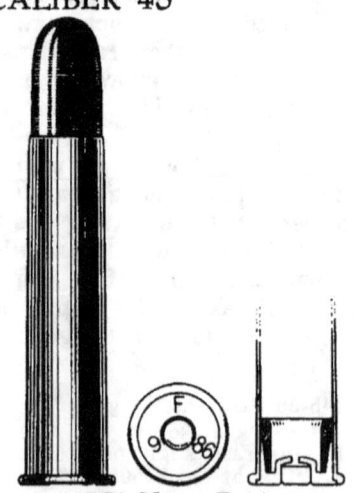

(45-70 Morse Pat.)
Total length 2²⁵⁄₃₂ in.

2 25/32 in. tinned case with detachable 2-piece head and copper primer
   Dia. at head .505
   Dia. at mouth .478
500 gr. round nose, lead bullet
70 grs. black powder

These cartridges, made under the Morse Patent of 1886, have no doubt been overlooked many times by collectors, since they so closely resemble the regular 45-70 Govt. The cross-section view illustrates the construction of the 2-piece head. They were made at the Frankford Arsenal in 1886 and 1887. The wording on an original box reads:

TWENTY
RIFLE CARTRIDGES
MORSE, MODEL 1886
GENERAL DIRECTIONS FOR RELOADING

Orders - Never reload except under the personal supervision of a competent officer.

Cautions - Never prime a loaded shell. Never load a primed shell without using a SAFETY-SOCKET

Manufactured at Frankford Arsenal

CALIBER 45

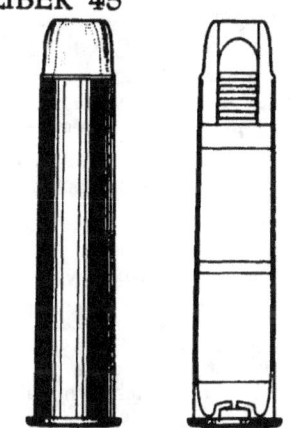

(45-70 RABBETHS PAT.)
Total length 2 15/32 in.
2 5/32 in. brass case
60 gr. lead bullet of .260 dia. enclosed in a wooden container
3 grs. black powder

7 grs. smokeless powder

The label just about describes this interesting cartridge which was written up in *Shooting and Fishing* in February of 1896.

20 45-70 RABBETH, SUB CALIBER 20
CENTRAL FIRE CARTRIDGE
Manufactured and Guaranteed
by the
WINCHESTER REPEATING ARMS
COMPANY
New Haven, Conn., USA

For use in all Rifles chambered for the 45-70 U. S. Govt. Cartridge

20 CARTRIDGES
  3 grs. black powder
  7 " smokeless powder
 60 " lead bullet

For Short Range Target Use
Especially adapted to indoor practice

CALIBER 45

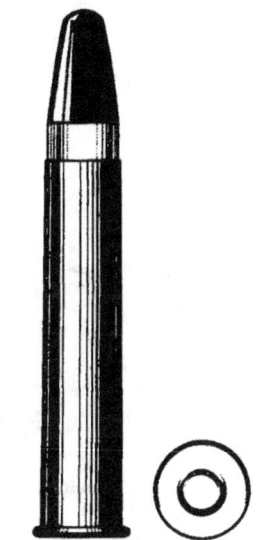

(45-78 Wolcott)
Total length 3 9/16 in.
2 9/16 in. brass case
   Dia. at head .505
   Dia. at mouth .476
475 gr. paper patched bullet
78 grs. powder

H. H. Wolcott, after whom these cartridges were named, was President of the Starr Arms Co. He was the inventor of the Wolcott carbine, Pat.

Nov. 27, 1866. It is thought that the name Wolcott was given to the bullets only and that they were loaded with various powder charges in .45 caliber cases. The label from an original box of the bullets reads:

```
              25
    .45 cal.    475 grs.
      WOLCOTT BULLETS
           Patched
      Manufactured by the
UNION METALLIC CARTRIDGE CO.
      Bridgeport, Conn., USA
```

Hand written on another old box is this description:

```
       10 C. F. Ctgs.
        45-70 WOLCOTT
          Loaded With
      M. H. DUPONT POWDER
```

CALIBER 45

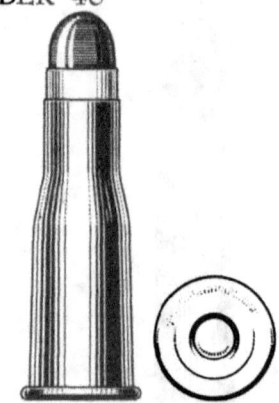

(45-85 Ward Burton Experimental)
Total length 2 11/32 in.
1 13/16 in. brass, necked case
  Dia. at head .657
  Dia. at mouth .483
400 gr. paper patched, lead bullet
85 grs. black powder

Used in the Ward-Burton Magazine Rifle in the experimental tests of 1872. It was first reported by the Frankford Arsenal Board on Nov. 27, 1871. W. G. Ward and B. Burton collaborated together to design this 4-hole Berdan primer cartridge. During the Centennial Exposition at Philadelphia in 1876, it was exhibited as cartridge No. 110. Manufactured by UMC, it is quite scarce indeed today.

CALIBER 45

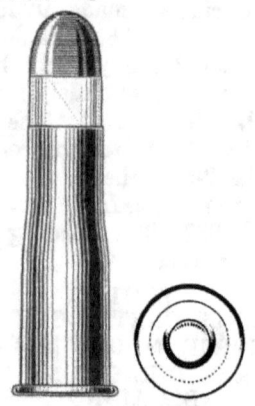

(45 Peabody-Martini Carbine)
Total length 2 5/16 in.
1 5/8 in. brass, necked case
  Dia. at head .581
  Dia. at mouth .477
405 gr. paper patched bullet
55 grs. black powder

This is the short cartridge for carbine use. A larger one, 45-85-480, was used in the rifle. The Peabody-Martini rifles and carbines were used as military arms by both England and Turkey. The cartridge shown here was the type used by the Turkish troops in the Russo-Turkish War.

CALIBER 45

(577-450 Martini-Henry)
Total length 3 1/8 in.
2 5/16 in. brass, necked case
  Dia. at head .650
  Dia. at mouth .492
480 gr. paper patched bullet
85 grs. black powder

This is the later solid drawn, brass case version of the earlier 577-450 Martini-Henry which was described in the preceding section, Center Fire, Part I.

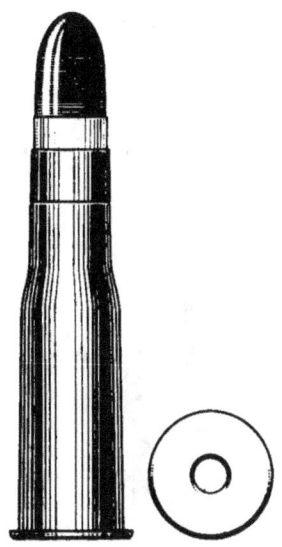

CALIBER 45

(45-100 Ballard Everlasting)
Total length nickeled case only
—2 13/16 in.
  Dia. at head .498
  Dia. at mouth .487
Uses Berdan primer

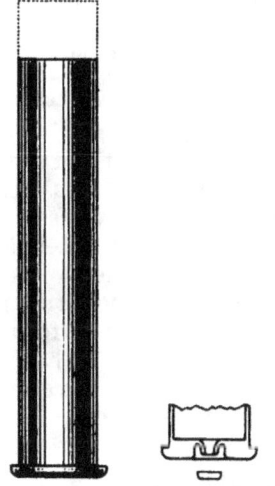

The 4 4/100 and 4 5/100 Everlasting cases are the same, with but one minor exception. In the latter, the mouth is counter-bored to take the .45 caliber bullet, while the former sometimes had a small raised ring at the mouth. These cases, which are quite scarce today in this caliber, were designed for continued reloading. They were advertised in the 1882 Ballard Catalog at the following prices, 4 4/100 - 9c each and the 4 5/100 - 10c each. Paper patched bullets for the cases were also advertised. The dotted lines on the one illustrated here indicate the original length. This particular case undoubtedly was cut down to fit a shorter chamber.

CALIBER 45

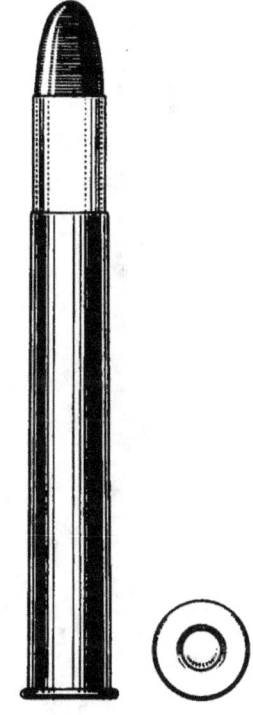

(45-120-550 Sharps)
Total length 4 5/32 in.
2 7/8 in. brass straight case
  Dia. at head .504
  Dia. at mouth .474
550 gr. paper patched, lead bullet
120 grs. powder

Used in the Sharps-Borchardt Express rifle which was introduced in 1879 and was discontinued two years later. It was designed for long range hunting on the Western Plains. The short life of the rifle was due to the closing of the Sharps Rifle Co. in October of 1881 for lack of capital.

Cartridges were made by Sharps and UMC. These large cartridges were for the most part loaded with 100 grains of powder. They may be found, however, with the 120 grain load . . . as was the specimen illustrated, which came out of an original box so labeled.

CALIBER 45

CALIBER 45

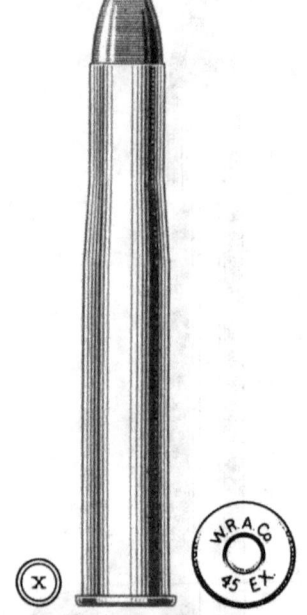

(45-125 Winchester Express)
Total length 3⅝ in.
3¼ in. brass, necked case
    Dia. at head .553
    Dia. at mouth .476
300 gr. express bullet
125 grs. powder

Used in the Winchester single-shot rifles which were made to order after October, 1886. These cartridges were also designed for heavy duty hunting of the late 1880's and early 1890's.

(45 - 3¼ Sharps)
Total length 4 3/16 in.
3¼ in. brass straight case
    Dia. at head .505
    Dia. at mouth .471
500 gr. paper patched bullet
120 grs. black powder

One of the longest cartridges produced in America. It was used in the Sharps-Borchardt Express rifle which was, upon order, chambered for this 3¼ in. case. At least one Sharps Long Range Creedmoor rifle is known to have been chambered for this same cartridge. These .45 Sharps in the 3¼ in. case are scarce collector's items of the present day.

CALIBER 450

(450-3¼ Straight)
Total length 3 13/16 in.
3¼ in. brass, straight case
270 gr. paper patch bullet with
    copper tube

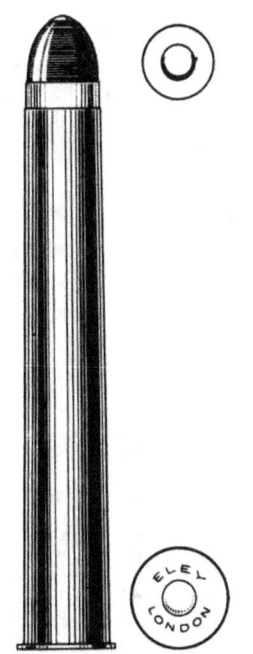

Another of the big game sporting cartridges produced in England by Eley of London.

CALIBER 450

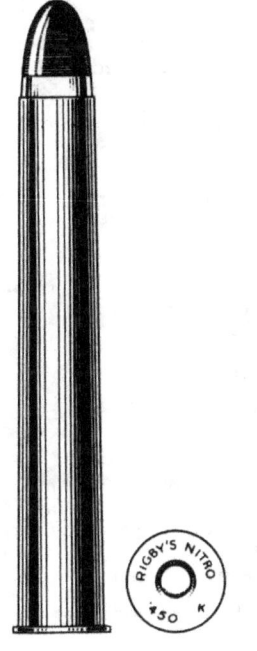

(450-3¼ Straight)
Total length 3¹³⁄₁₆ in.
3¼ in. brass straight case
400 gr. paper patched bullet

One of the English Kynoch big game cartridges.

CALIBER 455

(455 II Webley Revolver)
Total length 1¼ in.
¾ in. brass case, .478 dia.
265 gr. conical, lead bullet
7 grs. powder

The largest military revolver cartridge in use at the start of World War II was the 455 II Webley. It was one of the official revolver cartridges of the English Army. There was also a .455 rimless auto cartridge made for the Webley Automatic.

CALIBER 50

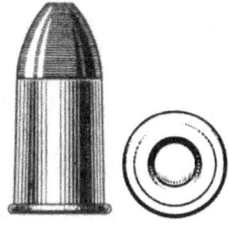

(50 Remington Navy)
Total length 1⁹⁄₃₂ in.
²⁷⁄₃₂ in. brass case with No. 2 primer
    Dia. at head .565
    Dia. at mouth .538
300 gr. flat nose, lead bullet
25 grs. black powder

One of the first metallic center-fire cartridges, with an outside primer, to be issued generally to U. S. troops was the .50 Remington. The Remington single shot pistol which used these cartridges was in use by the Service for only about ten years.

CALIBER 500

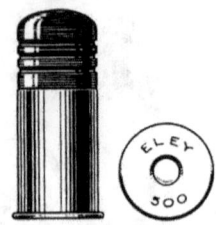

(500 Revolver and Carbine by Eley)
Total length 1¼ in.
$^{25}/_{32}$ in. brass case

No data seems available on this interesting cartridge produced by Eley of London.

CALIBER 50

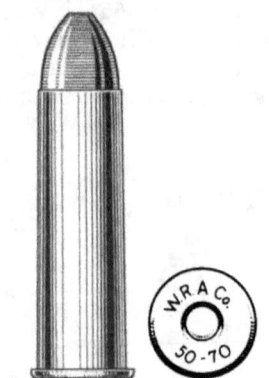

(50-70 Musket, also called .50 Govt.)
Total length 2¼ in.
1¾ in. brass case
    Dia. at head .565
    Dia. at mouth .533
450 gr. conical, flat nose, lead bullet
70 grs. black powder

This cartridge was still listed in the Winchester Catalog of December, 1896—even though most guns using this caliber were of earlier origin. Quite a few 50-70 rifles were issued to the service in the four years between 1869 and 1873. A shorter case length of this cartridge was used in the carbine.

CALIBER 50

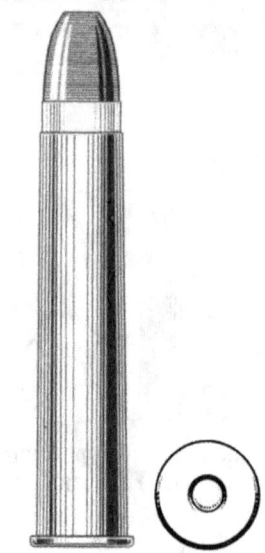

(50-90 Sharps)
Total length 3$^{7}/_{32}$ in.
2½ in. brass, straight case
    Dia. at head .567
    Dia. at mouth .528
473 grs. paper patched, lead bullet
90 grs. black powder

This cartridge, while a collector's item, can still be found on cartridge dealers' lists in limited quantities. It was used in the old Sharps B. L. sporting rifle for heavy game hunting in the West.

CALIBER 50

(50-105 Winchester Express)
Total length 2¾ in.
2⅜ in. brass case
    Dia. at head .551
    Dia. at mouth .531
300 lead, copper tube, express bullet
105 grs. powder

One of the rare cartridges to be found

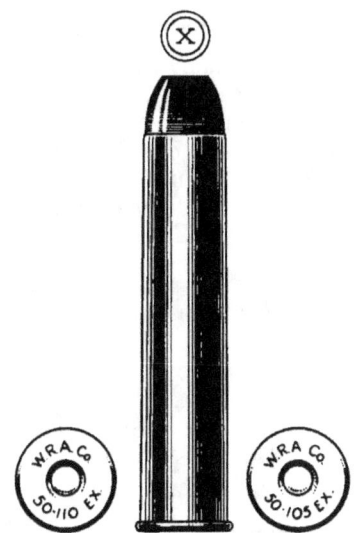

on some of the old Winchester cartridge boards. The Winchester Repeating Arms Company today can throw no light on this, obviously experimental, cartridge. Since the measurements are identical with the early 50-110, the head stamps of both cartridges are illustrated beside the case.

CALIBER 50

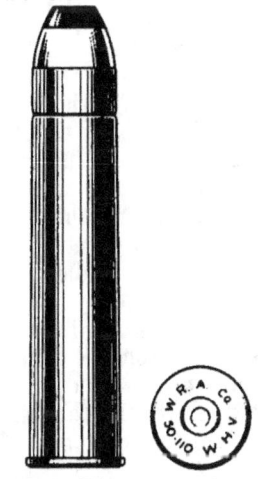

(50-110 Winchester High Velocity)
Total length 2¾ in.
2⅜ in. brass case

300 gr. metal jacket, soft point bullet
110 grs. powder

This cartridge was developed for use in the Model 1886 Winchester Repeating Rifle. This rifle came out first in the 45-70 caliber and shortly thereafter it was produced in the 50-110 caliber. Various types of bullets have been produced . . . lead, full jacket, soft point and hollow point.

CALIBER 50

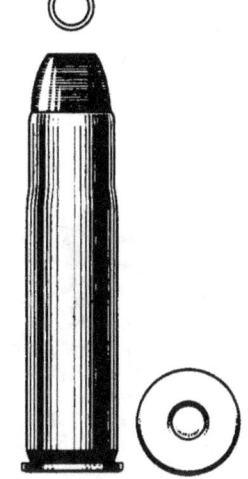

(50-115 Bullard)
Total length 2¹⁷⁄₃₂ in.
2⁵⁄₃₂ in. brass case—semi-rimless
   Dia. at head .583
   Dia. at mouth .542
300 gr. flat nose, copper tube, express bullet
115 grs. powder

This is one of the first, if not the first, semi-rimless cartridge produced in this country. It also was the first to have a "solid web" head, that is one in which the primer pocket does not project up into the case. The cartridge was first listed and described in the 1887 Bullard Catalog. They were used in both the Bullard single shot and the Bullard sporting magazine rifles.

CALIBER 50

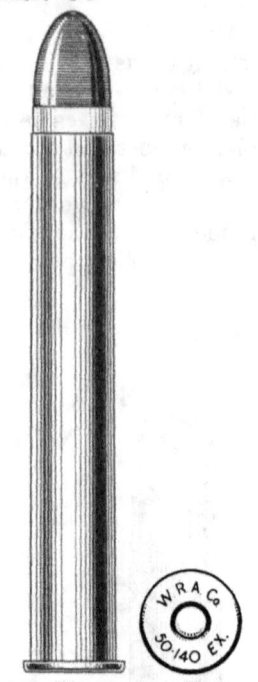

(50-140 Winchester Express)
Total length 3¹⁵⁄₁₆ in.
3¼ in. brass case
    Dia. at head .551
    Dia. at mouth .528
Paper patched bullet
140 grs. black powder

Another of the early experimental cartridges, the record of which seems to have been lost . . . at least the manufacturers have no data on it now. It is very similar to the big .50 Sharps cartridge . . . and even though there is a very slight difference in the miked measurements, it is possible that this was developed for use in the big .50 Sharps rifle and the Winchester single shot rifle.

CALIBER 50
  (50-3¼ Sharps)
  Total length 3¹⁵⁄₁₆ in.
  3¼ in. brass straight case
    Dia. at head .565
    Dia. at mouth .525
  473 gr. paper patched, lead bullet

140 grs. black powder

Not only among the rare, but one of the largest rifle cartridges ever produced in this country. The one illustrated here has the smaller bullet of 473 grs., which was the regular commercial load furnished by Winchester.

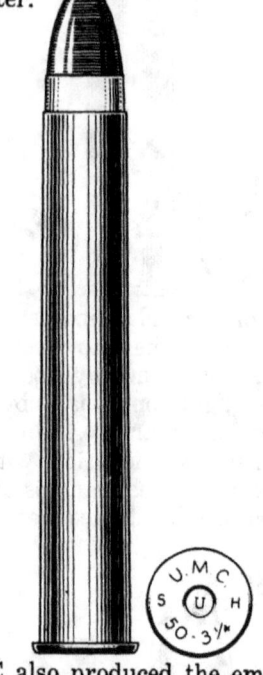

UMC also produced the empty cases in boxes of fifty for those who wished to load their own. In the Appendix—Bullet Section—is illustrated one of the big commercial 700 gr. paper patched bullets for this cartridge. Sharps did not produce this big cartridge and it is believed that any of the .50-90 heavy Sharps rifles using it were especially rechambered for its use. From all data available, there must have been only a few made due no doubt to the fact that it came out too late for the big buffalo hunting days.

CALIBER 500
  (500 Eley Express)
  Total length 3⁹⁄₁₆ in.
  3 in. coiled brass case with

iron head
  Dia. at head .579
  Dia. at mouth .520
340 gr. paper patched, copper tubed express bullet
120 grs. black powder

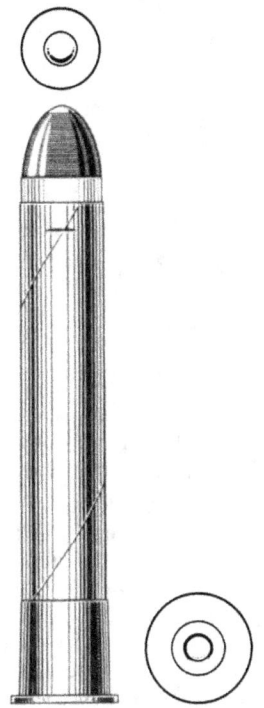

One of the early English sporting cartridges which is described by the label on a box of them as follows:

10 PATENT AMMUNITION
for Express Sporting Rifle

Expanding brass case which may be reloaded several times

ELEY BOXER
500 BORE
Powder 120 No. 1   Bullet 340 gr.
Eley Bros. Ltd.
London

CALIBER 500
(500-3¼ Express)
Total length 4 in.
3¼ in. brass straight case
  Dia. at head .575
  Dia. at mouth .530
Conical, lead bullet

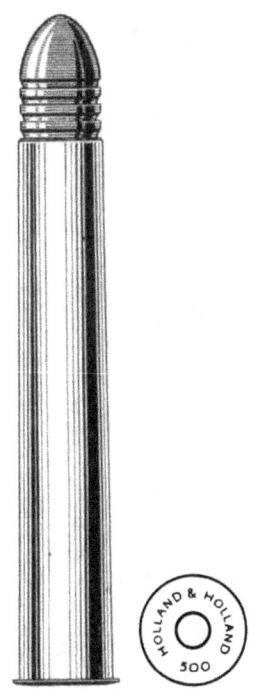

Another of the early, interesting big game cartridges produced by Holland & Holland, England. These cartridges were produced with three types of bullets . . . solid lead, as illustrated . . . hollow point . . . explosive head.

CALIBER 55

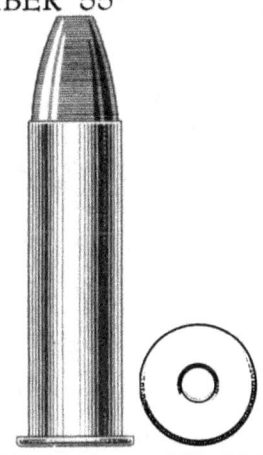

(55-100 Maynard Model 1882)
Total length 2⁹⁄₁₆ in.
1¹⁵⁄₁₆ in. brass case, .590 dia.

151

Conical, flat top, lead bullet
100 grs. black powder

Used in the Maynard Improved Hunters Rifle, Number 11 — this early sporting cartridge has long since become a collector's item.

CALIBER 57

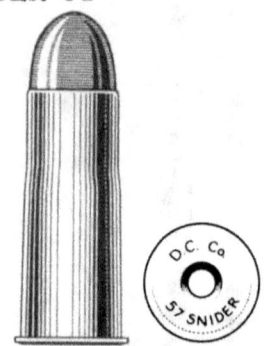

(57 Snider, also known as 577 Snider)
Total length 2⅛ in.
1¹⁹⁄₃₂ in. brass case with slight neck
    Dia. at head .659
    Dia. at mouth .600
480 gr. round nose, lead bullet
70 grs. black powder

A Canadian sporting cartridge, produced by Dominion Cartridge Company of Montreal.

CALIBER 58

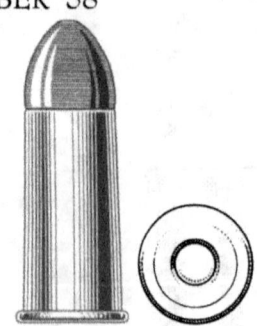

(58 Roberts)
Total length 1⅞ in.
1⁵⁄₁₆ in. brass case with Berdan type primer
620 gr. lead bullet
60 grs. black powder

Used in the Roberts B. L. Army Rifle ... an alteration of the Springfield and Enfield rifles. This cartridge was developed very shortly after these rifles were altered under the Roberts Patent of June 11, 1867.

CALIBER 58

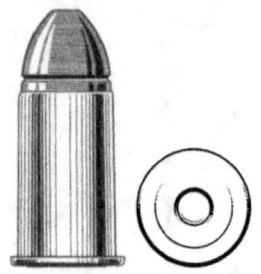

(58 Carbine)
Total length 1¹⁹⁄₃₂ in.
1³⁄₃₂ in. brass case
530 gr. flat nose, lead bullet
40 grs. black powder

This is the Berdan carbine cartridge named after Col. Hiram Berdan of the U. S. Sharp Shooters. He resigned his commission in the army to give his full time to perfecting a satisfactory breech-loading rifle which he hoped the United States would approve. His arms were produced by Colt Patent Fire Arms Mfg. Co. of Hartford, Conn., under Berdan Patent of March 30, 1869, in both rifle and carbine models.

CALIBER 58

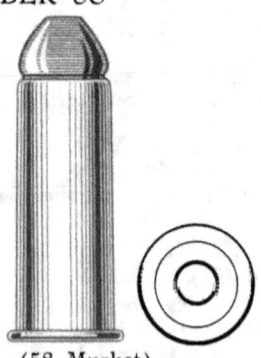

(58 Musket)
Total length 2¹⁄₁₆ in.

1⅝ in. brass case
    Dia. at head .642
    Dia. at mouth .615
530 gr. lead bullet
80 grs. black powder

The 58 Berdan Musket Cartridge for the Colt Berdan Musket is one of the largest center fire musket cartridges ever manufactured in this country. Like the carbine number, it was used in the military arms altered under Berdan's Patent No. 88,436.

CALIBER 58

CALIBER 60

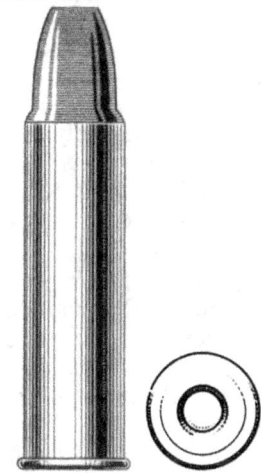

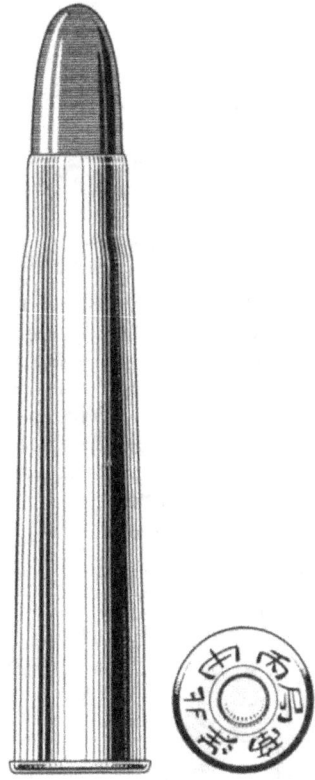

(58 Dangerfield & Lefever)
Total length $2^{25}/_{32}$ in.
$2^{3}/_{32}$ in. brass case, .635 dia.
Conical lead bullet

Used in the Dangerfield & Lefever rifle, a comparatively unknown arm patented Sept. 3, 1872. It will be noted that the case of this cartridge is very similar to the .58 caliber shot shell used at the time the .58 caliber Muskets were altered from muzzle-loading to breech-loading... and into shotguns. The cases were frequently used in the remodeled muskets which may account for the fact that Dangerfield & Lefever adopted it as the case for their big rifle barrel. The known specimen of this arm is also equipped with a 12 gauge shotgun barrel.

(60 Chinese Jingal)
Total length $4⅝$ in.
$3^{23}/_{32}$ in. brass, necked case
    Dia. at head .809
    Dia. at mouth .643
Cylindrical, round nose, lead bullet

The Chinese bolt action "Two Man Jingal" rifle made in China in 1892 is the real ancestor of the German anti-tank arm. Some of these captured arms were brought back as trophies by members of the International Forces during the Boxer Rebellion of 1900.

CALIBER 15 mm.

(15 mm. Revolver)
Total length $1^{15}/_{32}$ in.
$^{29}/_{32}$ in. brass case
    Dia. at head .610

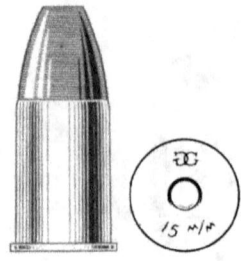

Dia. at mouth .606
Conical, flat nose, lead bullet

Produced by the French firm of Gevelot & Gaupillat, this is one of the largest center fire revolver cartridges known. Rarely do they show up on cartridge dealers' lists. What revolver handled this huge number is not known . . . at least no such arm has appeared on any American gun dealer's list in recent years.

CALIBER 70

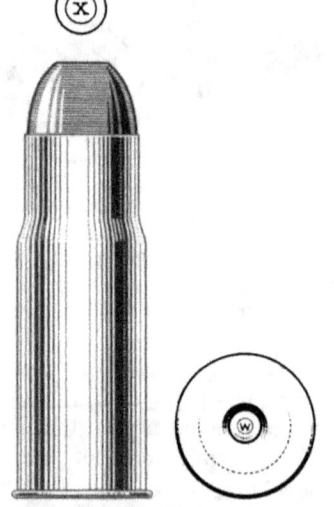

(70-150 Winchester Express)
Total length 2⅝ in.
2³⁄₁₆ in. brass, necked case
Dia. at head .803
Dia. at mouth .725
Copper tubed, lead bullet
150 grs. powder

Even though it was displayed on the early Winchester cartridge boards, virtually little if anything is known of this giant among cartridges. Apparently no one knows, for sure, what gun was chambered for it. It is fairly well agreed that it must have been a single shot since it seems inconceivable that any of the Winchester magazine rifles could handle such a large cartridge . . . unless it was a special model job produced only for exhibition.

The cartridge was exhibited at the Centennial Exposition in Philadelphia in 1876, but even the records of that Exposition are very elusive.

The records of the Winchester factory seem unable to throw any light upon this item. About the only thing which can be said about it is that it is a very desirable rarity of the first order . . . found only in a very few collections.

To Paul S. Foster, Chief Inspector of the Ammunition Division of the Winchester Repeating Arms Company, goes the credit for clearing up the mystery surrounding this giant. He writes in "The Gun Digest" (6th edition) that recently discovered factory tool records show this cartridge to have been produced in 1888. A special rifled barrel was fitted to one of their Model 87 Lever-action shotguns. He reports that one of the old employees of Winchester had told him of seeing, and puzzling over, such a gun in the old company's testing range. It will be recalled that the Model 87 was the first shotgun made by Winchester. A cartridge dealer some years back had two full boxes of these rarities . . . verifying the fact that not all of them appeared on cartridge boards made up by Winchester.

CALIBER 72

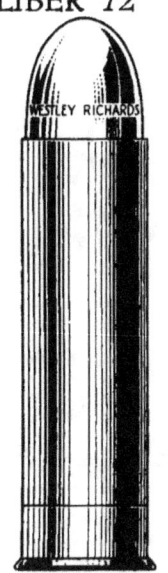

(72 Grouse Ejector)
Total length 3 5/16 in.
2 9/16 in. brass case

Dia. at head .802
Dia. at mouth .795
Brass jacketed bullet stamped
Westley Richards Patent
1897-01

Used in the Westley Richards, Faunetta and Paradox rifles for hunting wild boars, lions, tigers, and elephants in India. These rifles were of unusual construction. They were smooth-bore to within three or four inches of the end, then were rifled on to the muzzle. The only reason for the name seems to be that a 12 ga. shot shell case labeled "Grouse Ejector" was used for this large brass-jacket bullet.

shot

SEPARATE PRIMED

PIN FIRE

RIM FIRE

CENTER FIRE

SHOT

For nearly four hundred years, these little lead balls in varying sizes have been the sportsmen's friend. They were carried in shot pouches for the old muzzle-loaders, from the days of the wheel lock on down through the percussion era. Or, for added convenience they were put up in paper and wire containers. They were to be found in the forefront of every step of cartridge development. In fact shot shells and cartridges have just about run the gamut of types and calibers.

While there have been numerous improvements inwardly, the outside appearance of shotgun shells particularly have undergone very slight change in the past seventy-five years. So far as caliber or bore is concerned, the trend is back to smaller sizes . . . as it was during the wheel lock days when the .35 or .40 caliber was not unusual.

Many a lad was thrilled at the turn of the century to see some favorite western character break glass balls in the circus. What many of them did not know, at the time, was that shot cartridges were being used . . . both for safety to spectators and for efficiency in breaking the balls.

Of course shot cartridges were used for more than mere circus display. Many shooters will testify to this.

There is a greater variety of loading methods among the shot cartridges than among the shotgun shells. For instance some are loaded flush with the mouth of the case. several with pronounced crimps, many in a paper container, and not a few in a wood container similar to a bullet.

In many homes of by-gone years, one could find a box of brass cases for the loading or reloading of shotgun shells. But with the variety of loads in modern ammunition, the average sportsman has no need to "prepare" his own shot shells. Thus another chapter has been added to the growing list of Americana of the past.

Two English shot cartridges which enjoyed quite an extensive use in the 1850's were the Needham and the Lancaster.

While not drawn to scale, the two cross-section views will give a fair idea of their construction. They are included for their interest in the cartridge story.

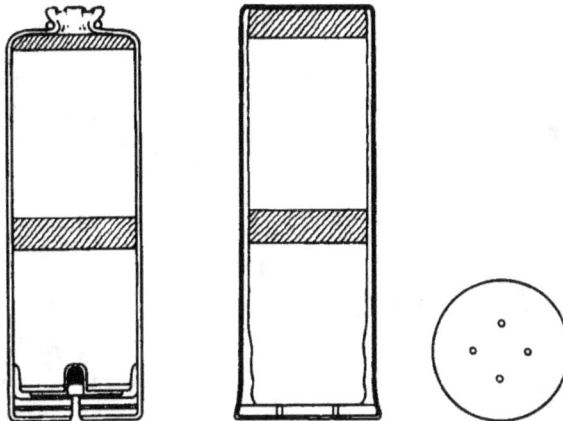

The Needham was of the combustible envelope type. Its base contained a percussion primer in the zinc wad. The paper case being consumed upon firing, the zinc wad was pushed forward in the chamber by the next cartridge. It then served as a front wad for that charge.

The Lancaster was composed of a thin, copper case and head. Within the head was an iron disk with four holes. The fulminate was placed between this disk and the head. Thus when the head of the cartridge received a sharp blow, the fulminate was detonated and the flame, passing through the holes, ignited the powder charge.

CALIBER 5 mm.

(5 mm. Shot R. F.)
Total length $27/32$ in.
Light orange paper case
Copper head

One of the smallest shot cartridges in use. This one was manufactured by Gustav Genshow and was brought back recently from Europe. The headstamp SAS of an Argentine cartridge is also shown beside it. Other than the headstamps, they are virtually identical. The paper case of the Genshow is orange while the case of Argentine number is red.

The letters SAS stand for Spreafico . . . a cartridge manufacturer at Buenos Aires. The full name is Spreafico, Vivda de Juan (widow of John Spreafico.)

CALIBER 22

(22 Long R. F.)
Total length $25/32$ in.
Copper case, rolled rim
Manufactured by Peters Cartridge Co.

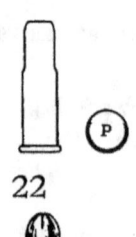

CALIBER 22

(22 Long Rifle R. F.)
Total length 31/32 in.
Brass, rose crimped case
Manufactured by Rem-UMC

CALIBER 32

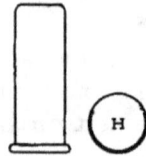

(32 Long R. F.)
Total length 7/8 in.
Copper case with rolled rim
Manufactured by WRA

CALIBER 32

(32 Long R. F.)
Total length 1 7/32 in.
25/32 in. copper case with wooden container
Manufactured by Rem.-UMC

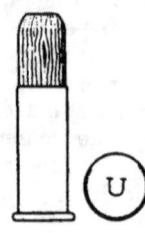

CALIBER 32

(32-20 Shot C. F.)
Total length 1 5/16 in.
1 5/16 in. brass, necked case with wooden container
Manufactured by Rem.-UMC

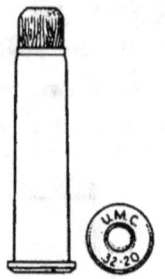

CALIBER 9 mm.

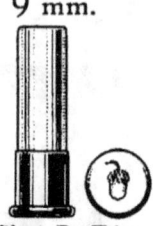

(9 mm. Shot R. F.)
Total length 1 1/8 in.
Orange colored, paper case
Copper base
Manufactured by Genshow

CALIBER 9 mm.

(9 mm. Winchester R. F.)
Total length 1 9/16 in.
Red paper case with copper head
Manufactured by WRA

Used in the Winchester Model 36 Bolt Action Shotgun. Foreign and domestic manufacturers used many colors in the manufacture of these cartridges . . . red, green, tan, etc.

CALIBER 38

(38-40 Shot C. F.)
Total length 1 19/32 in.
1 5/16 in. brass, necked case with paper container
Manufactured by Dominion

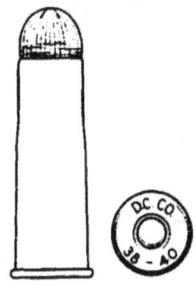

Cartridge Co.

CALIBER 41

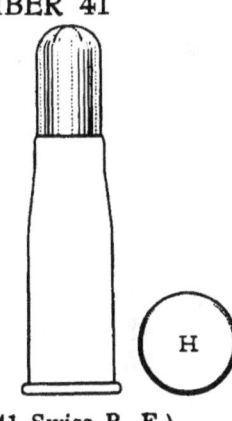

(41 Swiss R. F.)
Total length 2 3/16 in.
1 17/32 in. copper, necked case with dark blue or red paper container

CALIBER 11 mm.

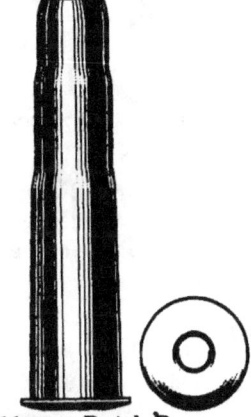

(11 mm. Dutch Beaumont C. F.)
Total length 2 7/16 in.
Brass, necked case—with wax wad.

Used in the Dutch Beaumont Military Rifle Model 1871.

CALIBER 44

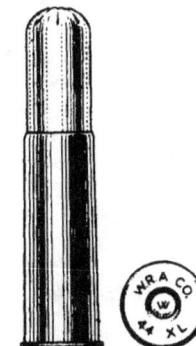

(44 XL - CF)
Total length 2 1/32 in.
1 9/32 in. brass case with red paper container
Manufactured by WRA

CALIBER 45

(45 Colt ACP)
Total length 1 1/4 in.
Brass, necked case with purple paper wad
Manufactured by Remington Arms

World War II issue. Survival kits for paratroopers included these shot shells, which were to be used primarily for small game.

CALIBER 45

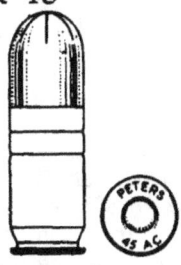

(45 Thompson)

Total length 1 7/16 in.
7/8 in. brass case with red paper container
Manufactured by Peters Cartridge Co.

This is a commercial cartridge for guard or riot purposes. Used in the Thompson Sub-Machine Gun.

CALIBER 410

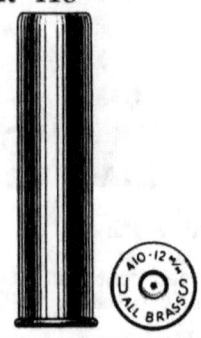

(410 Gauge ALL BRASS)
Total length 1 7/8 in.
Brass case with paper wad
Manufactured by U. S. Cartridge

CALIBER 50

(50 Remington Navy Pistol)
Total length 1 9/32 in.
7/8 in. brass case with wooden container
Manufactured by Rem.-UMC

Cross-section illustrates the construction of wooden container. Used in the single shot Remington pistols.

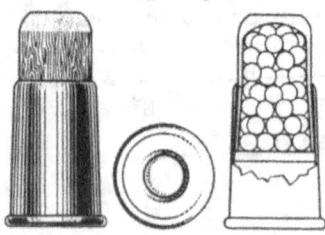

CALIBER 55

Many Maynard rifles were cased in a leather or wooden box with an extra barrel or two of a different caliber. For instance one such case contained a .35 caliber barrel, a .40 caliber barrel and a barrel chambered for the .55 caliber shot shell.

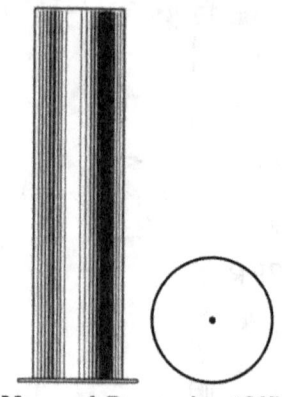

(55 Maynard Percussion 1865)
2 1/4 in. brass case

CALIBER 55

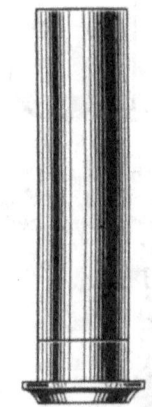

(55 Maynard Model 1873 CF)
2 3/8 in. brass case

Used in the Maynard B L shotgun No. 1 or No. 2, Model 1873. No. 1 had a 26 in. barrel and No. 2 a 36 in. barrel.

CALIBER 55

(55 Maynard Model 1882 CF)
2 5/16 in. brass case

Used in either the No. 1 or No. 2 Maynard B. L. shotgun, Model 1882.

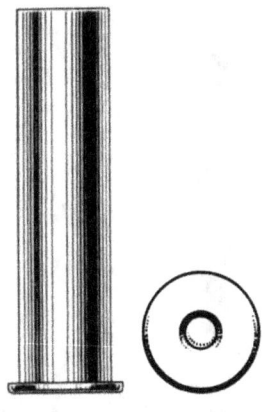

CALIBER 32

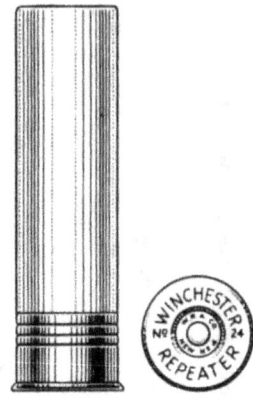

CALIBER Gauge 20

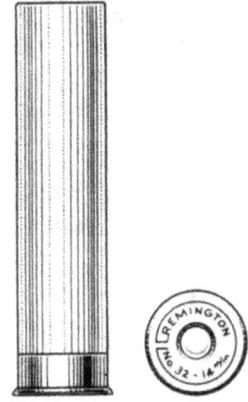

(32 Ga. Remington—14 mm.)
Total length 2⅜ in.
Red paper case, brass head.

CALIBER Gauge 24

(24 Ga. Winchester Repeater)
Total length 2⁵⁄₁₆ in.
Red paper case, brass head

(20 Ga. Bouron)
2²⁹⁄₃₂ in. steel case

One of the most interesting shot shells encountered—P. Bouron was a gun maker of New Orleans, La., during the early cartridge period. Here then is one of the early transition period cartridges. An auxiliary chamber for a percussion arm—still retaining the percussion cap—but providing the owner with the convenience of extra "cartridges" for the hunt.

T. L. Sturtevant on June 12, 1866, secured a patent (No. 55,552) for a similar type of cartridge. Whether the one illustrated was produced under this patent is not known.

CALIBER Gauge 20

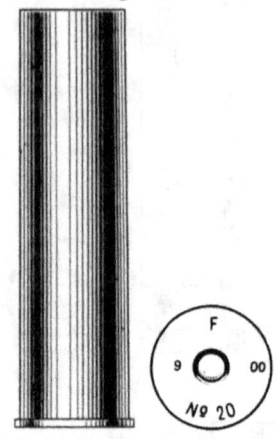

(20 Ga. F. A.)
2 17/32 in. brass case (also found in tinned case)

Manufactured at Frankford Arsenal ... undoubtedly for use as a military or guard load. It is said that the Army issued one single shot gun to each unit in service in Kansas and Nebraska. These were used for hunting grouse, prairie chicken, etc., to supplement the Army's rations.

CALIBER 64

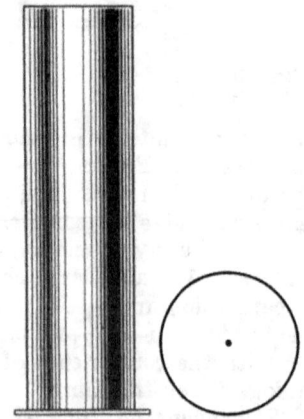

64 Maynard Percussion, 1865
2 15/32 in. brass case

These brass shot cases were used over and over again for reloading.

CALIBER 64

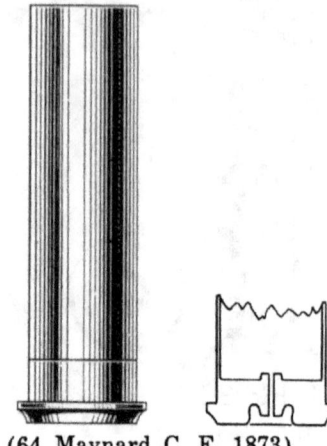

(64 Maynard C. F. 1873)
2 17/32 in. brass case

Used in the No. 3 and No. 4 Maynard B. L. shotgun. No. 3 had a 26 in. barrel while No. 4 was equipped with either a 28, 30 or 32 inch barrel. The cross-section shows the center flash hole and the anvil for the primer.

CALIBER 64

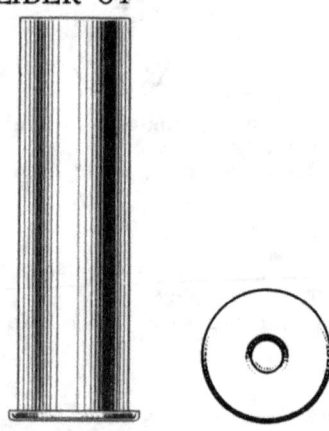

(64 Maynard CF 1882)
2 7/16 in. brass case

This shot shell came out of a Maynard-cased set which included, besides the shotgun barrel, a rifle barrel chambered for the 40-40 rifle cartridge.

CALIBER Gauge 16

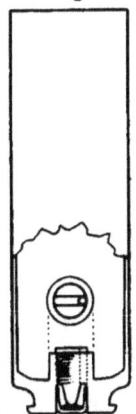

(16 Ga. Roper)
2⁷⁄₁₆ in. steel case

Used in the Roper 4-shot Revolving shotgun, made by the Roper Repeating Rifle Co., Amherst, Mass., Patented April 10, 1866. Since the exteriors of the Roper cartridges are very similar, only the detail view is shown. This one has a removable "anvil screw." Running the full length of the screw, at one edge, is the flash hole to carry the flame from the cap to the powder charge.

CALIBER Gauge 16

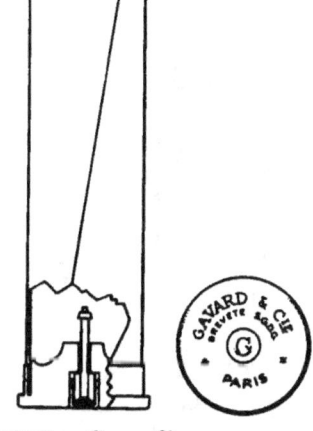

(16 Ga. Gavard)
2½ in. coiled steel case with brass head

A very interesting reloadable shot shell is this item of French manufacture. The cross-section view illustrates the anvil pin which also serves as an extractor for the percussion cap. This principle of a combination anvil and extracting pin was employed in a few cartridges of this country around 1870.

A similar cartridge to this was patented by A.N.C. Gavard, of Paris, in this country on April 19, 1870 (No. 102,109).

CALIBER Gauge 16

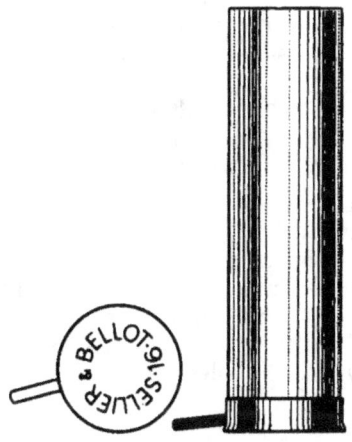

(16 Ga. Pin Fire)
2⁹⁄₁₆ in. paper case with brass head
Manufactured by Sellier & Bellot

Pin fire shot shells were made in a wide variety of calibers and colored paper cases.

CALIBER Gauge 14

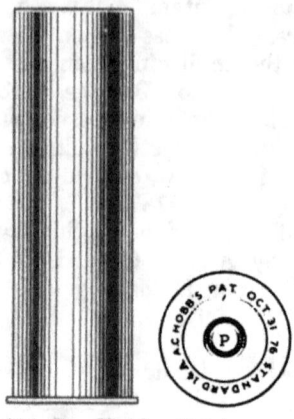

(14 Ga. Hobbs Pat.)
2¹¹⁄₃₂ in. brass case

CALIBER Gauge 12
(12 Ga. Roper)
2⅜ in. steel case

Two types of the Roper 12 ga. are illustrated. *Fig. 1* is the earlier type with the single flash hole and the anvil formed out of the solid head. *Fig. 2*, Roper with primer extractor pin which also serves as an anvil. Four flash holes were provided to convey the flame from the primer to the powder charge.

On April 17, 1866, Thomas L. Sturtevant of Boston secured a patent (No. 54,038) for a cartridge having such a primer extractor pin. The 4-shot Roper revolving shotguns using these 12 ga. shells were made by the Roper Sporting Arms Co., Hartford, Conn. They were in use up until 1880 or perhaps a little later.

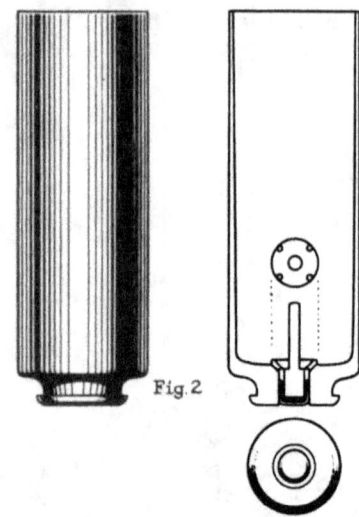

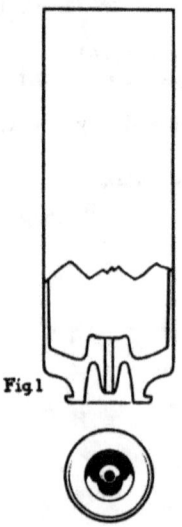

CALIBER Gauge 12

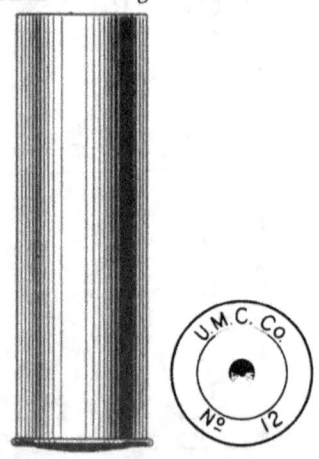

(12 Ga. UMC Percussion)
2⅝ in. brass case
Manufactured by UMC

166

Used in shotguns employing a percussion cap as a priming agent.

CALIBER Gauge 12
(12 Ga. Allen)
2¹³⁄₃₂ in. steel case

Ethan Allen secured a patent for this cartridge on May 16, 1865, (No. 47,688). The patent papers give this description, "Brass shell with steel base brazed inside. Front of shell

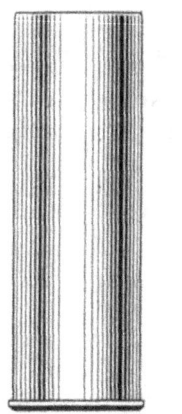

screw-threaded or roughened internally."
The cartridge used the Allen's Pat. primer shown below the head.

CALIBER Gauge 12

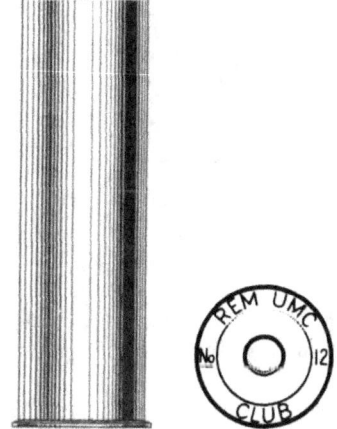

(12 Ga. Rem.-UMC Club)
2¹⁹⁄₃₂ in. brass case

This is one out of a wooden container holding fifty of these brass reloading cases. Many a winter evening in the old days was spent reloading a group of cases for the next hunt.

CALIBER Gauge 12
(12 Ga. UMC)
4 in. brass case

This lengthy number was also made in a 10 gauge, though what gun used either is at present unknown.

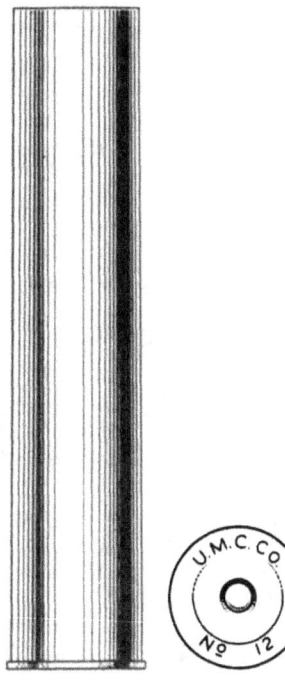

CALIBER Gauge 10

Though this cartridge is known as the Draper, it was actually patented by W. H. Wills (No. 45,292). It was described at the time as "Metallic, screw cap on base." It employs an ordinary percussion cap as is used on percussion revolvers.

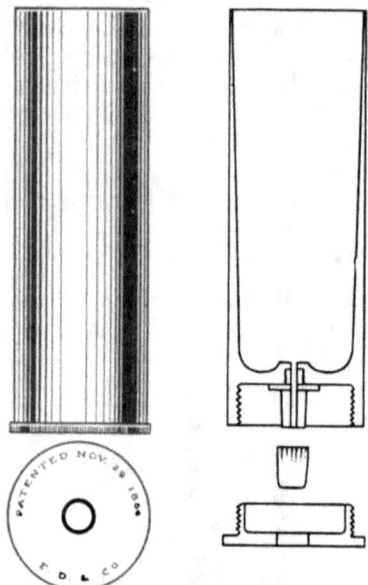

(10 Ga. Draper)
2⁹⁄₁₆ in. brass case
Manufactured by Draper & Co.

CALIBER Gauge 10

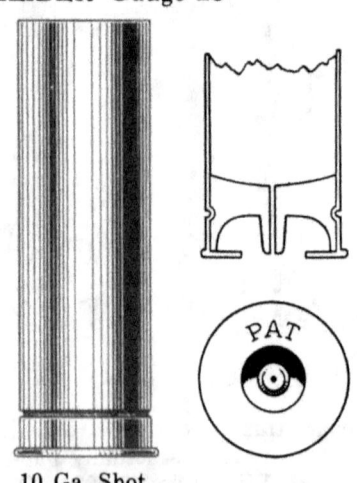

10 Ga. Shot
2⅝ in. brass

W. Tibbals of South Coventry, Conn., secured a patent (No. 51,243) on Nov. 28, 1865, for a cartridge having an . . . "internal anvil with nipple rests loosely against perforated base of shell." It would appear that this case, labeled only "Pat." is this very same Tibbals patent from a comparison with the patent drawing.

CALIBER Gauge 10

Caliber 10 Gauge
(10 Ga. Conical Base, Shot Shell)
2¹¹⁄₁₆ in. brass case

Reloading shells used in the 1880's. A label from an old box reads:

25   The   No. 10
CONICAL BASE
BRASS SHOT SHELLS
Patented March 3, 1885
Length   Inches
Manufactured by
THE AMERICAN BUCKLE & CARTRIDGE CO.
West Haven, Conn.   U.S.A.
THE BEST BRASS SHOT SHELL MADE

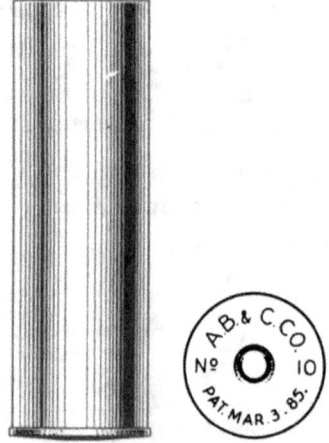

CALIBER Gauge 10

Such a cartridge was patented Jan. 18, 1870, No. 98,995, by Sewell Newhouse of Oneida, N. Y. His patent provided for a cartridge case having a collar in the center of the head, perforated with two or more holes, and having a movable anvil which

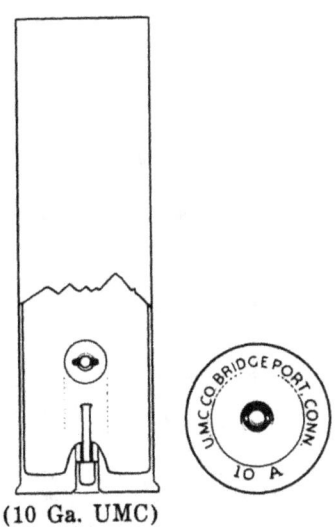

(10 Ga. UMC)
2⅞ in. brass case

served as a cap extractor.

CALIBER Gauge 8

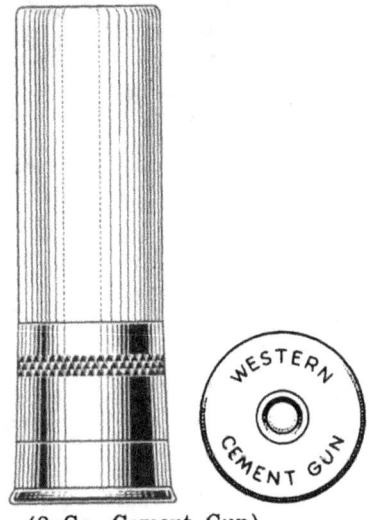

(8 Ga. Cement Gun)
Total length 3 in.
Red paper case with brass head

Remington Arms introduced a device known as the "Cement Kiln Gun." It was developed for the purpose of breaking up clinker rings in a cement kiln by means of a heavy cylindrical lead slug. The gun was 8 gauge, cylinder bored. This is an example of how a firearm can be used to commercial advantage. The January, 1926, issue of *Scientific American* contained a detailed description of this gun, written by Captain E. C. Crossman. The article was entitled "Shooting Clinker Rings from Kilns."

CALIBER Gauge 4

(4 Ga. Rem.-UMC)
4 in. brass case

This large item was also made with a brass head and a paper case—the one by Winchester having a black colored case.

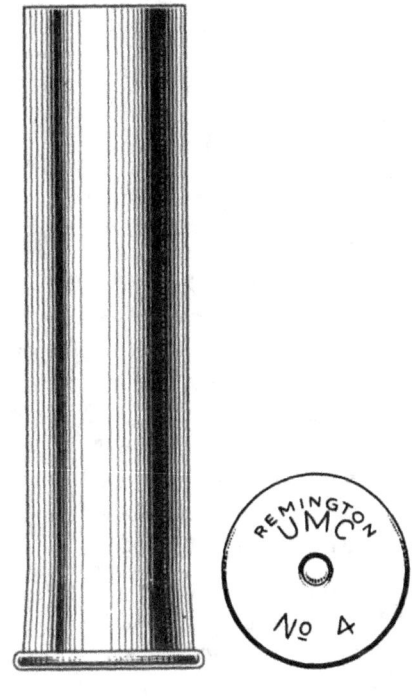

CALIBER Gauge 4

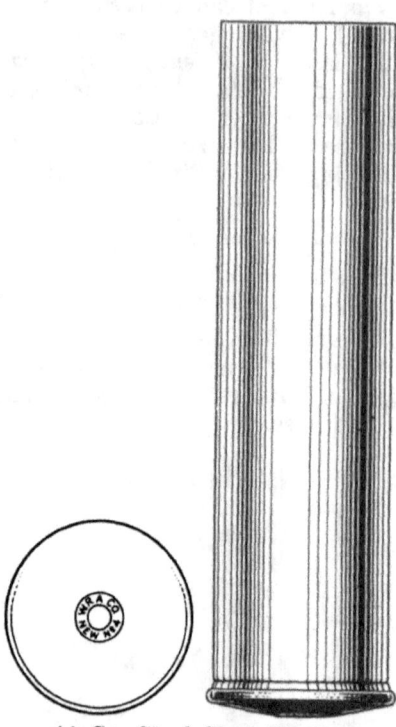

(4 Ga. Steel Shot Shell)
3$^{15}/_{16}$ in. steel case

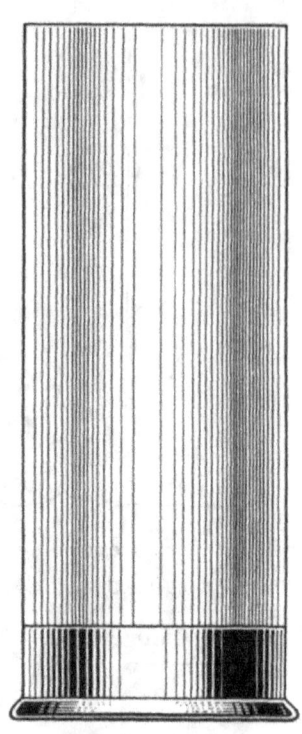

(2 Ga. Shot Shell)
Total length 3$^{15}/_{16}$ in.
Buff colored paper case with brass head

The specimen examined was equipped with a WRA New No. 4 Primer ... though this may have been from one of the numerous reloads this shell had experienced.

CALIBER Gauge 2

It is said that in other years these mammoth shells were used in swivel mounted guns on the sides of boats. They were used in the commercial hunting of wild game, ducks, geese, etc.

Another of the mammoth shot shells was the 3 gauge brass base, with paper case, put out by UMC.

There is also the giant among shot shells ... the 1 gauge ... specimens of which are to be found only in a few collections.

STANDARD SHOT SIZES
Standard Sizes, Soft and Chilled

No.	Chilled Shot No. in Oz.	Soft Shot No. in Oz.	Diameter in Inches	Diameter in Millimeters	No.	Soft Shot No. in Oz.	Diameter in Inches	Diameter in Millimeters
Dust		4565	.04	1.02	B	59	.17	4.32
12	2385	2326	.05	1.27	Air Rifle	55	.17½	4.44
11	1380	1346	.06	1.52	BB	50	.18	4.57
10	868	848	.07	1.78	BBB	42	.19	4.83
9	585	568	.08	2.03	T	36	.20	5.08
8	409	399	.09	2.28	TT	31	.21	5.33
7½	345	338	.09½	2.41	F	27	.22	5.59
7	299	291	.10	2.54	FF	24	.23	5.84
6	223	218	.11	2.79				
5	172	168	.12	3.02				
4	136	132	.13	3.30				
3	109	106	.14	3.53				
2	88	86	.15	3.78				
1	73	71	.16	4.06				

BUCKSHOT

Eastern Size	Western Size	Diameter in Inches	Diameter in Millimeters	Approximate No. in Lb.
4		.24	6.09	341
3	8 or 9	.25	6.35	299
2	7	.27	6.86	238
1	5 or 6	.30	7.62	175
0	4	.32	8.13	144
00	3	.34	8.64	122
000	2	.36	9.14	103

SHOTGUN GAUGES

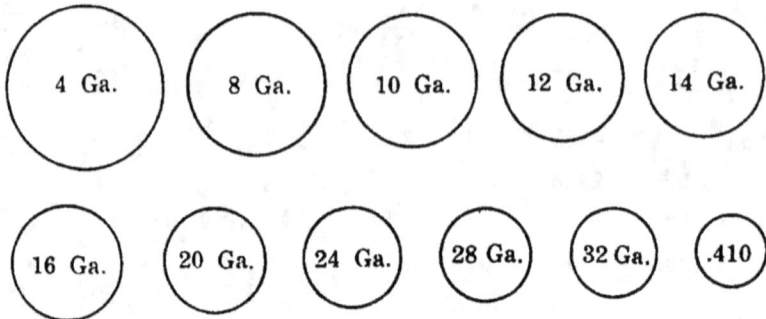

Gauge sizes in shotguns had a very interesting yet simple origin. In the dim past someone conceived the idea of designating the size of a shotgun bore from the number of spherical lead balls of that same diameter which could be cast from a pound of lead. The diameter of a lead ball weighing one pound was called a one gauge. This same ball divided equally into four balls of the same diameter gave the size for the four gauge. That pound ball broken down into twelve balls of equal weight and diameter gave us our popular twelve gauge size — and so on for the various gauges.

One exception to this common designation of gauges is the present day favorite. small gauge .410 shotgun. Its bore diameter is given in thousandths of an inch and this designation has been applied to both gun and cartridge.

Chamber diameters are larger than the bore diameter in each instance. If the barrel is choked, the diameter at the muzzle will be a bit smaller than that of the bore . . . depending upon the amount of choke.

Blank

PIN FIRE

RIM FIRE

CENTER FIRE

BLANK

Ordnance Memo No. 14 gives a description of the .50 caliber service blank. It is of the inside cup primed variety and is thought to be one of the first, if not the first, metallic blank center fire cartridges issued to the service.

Since then blanks have been produced in many calibers. Specimens included here range from the miniature 2 mm. pin and rim fires on through to the large .58 caliber rim fire. Various types of cases are shown.

Primarily blanks were developed for celebration, demonstration and display purposes. Since they contain no bullet, their only danger would be from the wad or powder burns at close range.

Included in this section is a specimen of a grenade launcher. This is similar to a blank . . . except that it has a wooden bullet in place of the paper wad. One of the most unusual is the "movie load" a blank which is especially designed to make a lot of noise with a great deal of smoke and flash.

Two other blanks were the circus load (known as the one-half charge blank) and the safety blank, with a minimum load.

CALIBER 2 mm.

(2 mm. Pin Fire)
Total length ⁵⁄₃₂ in.
Copper case

Smallest pin fire blank cartridge known. Used in the miniature watch charm single shot pistols.

CALIBER 2 mm.

(2 mm. Rim Fire)
Total length ³⁄₃₂ in.
Copper case

Produced mainly in Germany and Austria prior to World War II. Used in the miniature watch charm pistols and revolvers. The label from a box of these miniature blank cartridges reads as follows:

```
        20
     Randzünd
     Patronen
     Kal. 2mm.
```

CALIBER 4 mm.

(4 mm. Kolibri - RF)
Total length ¼ in.
Copper case

Used in the miniature blank alarm or starter pistols. This cartridge was manufactured by Genschow of Germany.

CALIBER 22

(22 Blank)
Total length 1³⁄₃₂ in.
Copper case with rose crimp
Manufactured by WRA

CALIBER 6 mm.

(6 mm. Flobert - RF)
Total length ⁷⁄₃₂ in.
Copper case

These little blanks were used in the German alarm and starter pistols. The Em-Ge gun used a flat horizontal bar with six chambers rather than a cylinder. The cartridge illustrated was made by Hirtenberger of Austria.

CALIBER 6 mm.

(6 mm. Perfumed - RF)
Total length ⁷⁄₃₂ in.
Copper case

Yes—a cartridge which dispenses perfume instead of a bullet. It must have been quite a favorite around ye old stables, hog pens, etc. The cartridge illustrated came in Hyacinth, and the head is colored a light magenta. Others could be supplied in Rose, Lilac, Violet and Lily of the Valley. The label from a box reads:

```
        10 Stück
    Parfüm - Patronen
      Caliber 6 m m
    Marke "DEMIMORS"
        Hyazinth
```

CALIBER 30

(30 Krag)
Total length 3¹⁄₁₆ in.
Brass full length case, rimmed
65 grs. black powder

The report of the Secretary of War, 1896, pictures and describes this cartridge. It is referred to as "The Rifle and Carbine Blank Cartridge, Whole Case, Caliber .30, Model 1893." The report also mentioned that this cartridge was being superseded by the paper ball blank, Model 1896. They

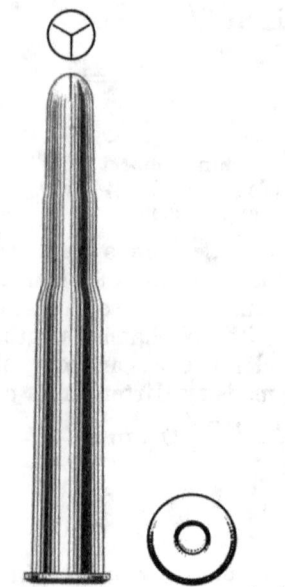

do not appear today on any dealer's list.

CALIBER 30

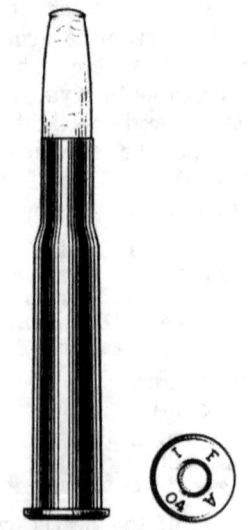

(30 Krag)
Total length 3 1/16 in.
2 9/16 in. brass case, rimmed
Paper bullet filled with a fine white powder

Used in the Krag-Jorgensen rifles and carbines.

CALIBER 30

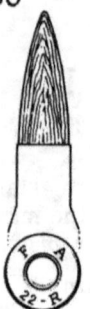

(30-06 Grenade Launcher)
Total length 3 5/16 in.
2 15/32 in. brass, necked case
10 gr. wooden bullet

The label from a box of these cartridges reads:

20
Cal. .30 Special Type Blank Cartridges
For use only with V.B. Grenades

FOR DEMONSTRATION PURPOSES
55 Grains 1 MR No. 18 and 5 grs. FFG
F. A. Primer No. 70
RIFLE CARTRIDGE CASE
10 GR. HARDWOOD BULLET
F. A. Drawing B-8022
Mfd. Frankford Arsenal Oct. 1922

CALIBER 30

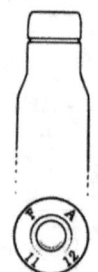

(30-06 FA Blank)
Total length 2 1/2 in.
Brass case with wax wad
Frankford Arsenal Nov. 1912

CALIBER 32

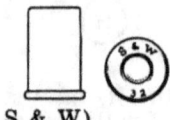

(32 S & W)
Total length 9/16 in.

CALIBER 38

Brass case ... rolled rim
Manufactured by the International Cartridge Co., Germany

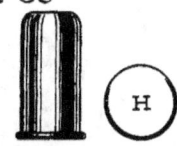

(38 R. F. Blank)
Total length 3/4 in.
Copper case
Manufactured by WRA

CALIBER (5 in 1)

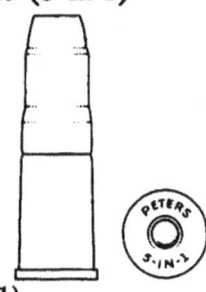

(5 in 1)
Total length 1 9/16 in.
Brass, double necked case
Manufactured by Peters Cartridge Co.

This blank cartridge is known as the "5 in 1" due to the fact that it can be' used in the following arms: 38-40 and 44-40 caliber rifles, 38-40, 44-40 and 45 caliber revolvers. It was developed as a movie load where lots of smoke and a loud report was desired.

CALIBER 44

(44 WCF or 44-40 blank for rifle)
Total length 1 9/16 in.
Brass case with rolled rim
Manufactured by WRA

How many such cartridges were used in the old 44-40 Winchesters in the Wild West Shows of bygone years may never be known ... but the total amount must have reached a staggering figure.

CALIBER 45

(45 FA Blank)
Total length 1 1/32 in.
Brass, necked case
Manufactured at Frankford Arsenal

Used in the 1909 Double Action Revolvers.

CALIBER 45

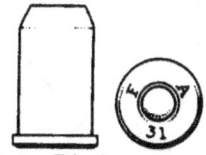

(45 Auto Rim)
Total length 27/32 in.
Brass case, beveled mouth

Used in the 1917 service revolvers using 45 Auto cartridges in a clip. These were developed for use without the clip.

CALIBER 45

(45-70 Govt.)
Total length 2 1/8 in.

Brass case with rounded cannelure
Manufactured by WRA

CALIBER 45

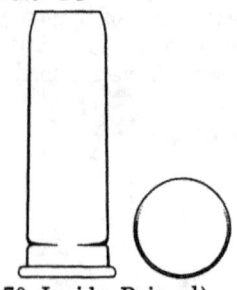

(45-70 Inside Primed)
Total length 1 19/32 in.
Copper case, beveled rim

Benet type inside cup primed.

CALIBER 50

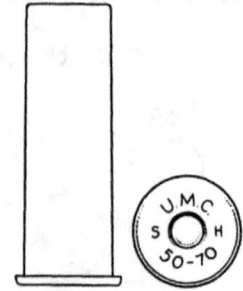

(50-70 Govt.)
Total length 1 11/16 in.
Brass case, rolled crimp with white paper wad
Manufactured by UMC

CALIBER 56

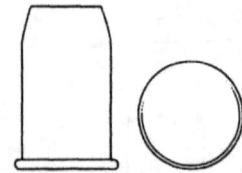

(56 Spencer RF)
Total length 31/32 in.
Copper case with beveled crimp

Used in the old Spencer repeating rifles and carbines.

CALIBER 58
(58 Roberts)
Total length 1 11/32 in.

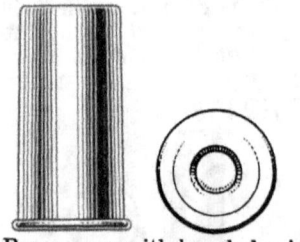

Brass case with beveled crimp
Manufactured by UMC

Used in the Roberts rifle and Roberts breech-loading army carbine. Roberts was a West Point graduate who developed a system of converting the old Civil War muzzle-loaders to breech-loading cartridge arms.

appendix

PRIMERS

BULLETS

LIST OF PATENTS (TO 1878)

MANUFACTURER'S HEADSTAMPS

CARTRIDGES OF THE '90'S

COMPARATIVE DIMENSIONS

DWM MARKINGS

DETONATING THE POWDER CHARGE

One of the most interesting phases of the whole story of cartridges is to be found in the detonation of the powder charge. The evolution of the modern primer from the burning stick required some five centuries filled with romance, history and science. Illustrated here are but a few of the steps in this pageant of progress.

Flame from a burning stick applied to priming powder scattered around the touch hole was the first means of detonation.

A burning cord held in a serpentine was used in the Matchlock period to ignite the powder charge.

Pyrite (fool's gold) brought into contact with a revolving wheel to throw off sparks was used in the early wheel locks.

Flint striking against steel provided sparks for detonating the powder charge in the days of the flintlock.

Tiny pellets made of fulminate of mercury were used in the pill lock arms. When struck by the hammer, the resulting sparks of flame ignited the main powder charge.

THE PRIMER

One of the finest descriptions of the primer ever written is to be found in an article in the *ARCA Arms Review* (1934) by the late Paul B. Jenkins, entitled "The Story of the Primer." Parts of it are used with the kind permission of F. Theodore Dexter, the Publisher.

"There are more of the combined Romances of History and of Science bottled up in any modern cartridge than in any other thing of its size in all the world . . .

Whatever cartridge you choose, its projectile—bullet, shot-charge or armor-piercing s h e l l — is unfailingly amazing in embodied capabilities of range, accuracy and energy. Its mechanically measured explosive is uniform with all its kind to within a few grains. Its case is as bright and as carefully turned out as a new government coin. Weatherproof against the elements, it is almost equally unaffected by storage through the years. In short, any modern cartridge is one of the most marvelous products of present-day science.

But of all its wonders and perfections, how many of us have ever adequately considered the primer set in the center of its head—in rim-fires its equivalent inside the rim—inconspicuously yet indispensably awaiting its summons to instantaneous efficiency? How innumerably many times life itself in some form hangs upon the instant sureness of response of those tiny cups and their magic contents! Failure there, and the whole weapon, though the finest product of half a dozen sciences combined, is no better than the "skull-cracker" of the Indian or the war-club of the South Seas savage.

How many of us have ever thought of the instantaneous chemical ebullition, the miniature cataclysm of violent action and reaction of molecules and their imprisoned forces, that goes on in that shiny drop of silvery material when the firing-pin wakes its quiesence to a Liliputian Vesuvius of erupting gas and flame? Yet a century and a half of ceaseless study and experiment in the laboratories of some of the world's foremost chemists have been devoted to effecting both the normal safety and harmlessness of that magic bead and its equally normal and certain waking in a tiny flaming fury of transformation at the sharp rap of the trigger-released striker on its thin metal shell.

Made by the millions daily in great factories, handled by all of us with no more thought of danger than if it were a bit of wood or stone, the process of blending its components is so fraught with possibilities of shattering destruction that the laws of empires have stepped in to order that but one man at a time may preside at the fuming witches' brew of the kettle in which its inert materials are united to produce that gray powder of infinitesimal crystals which alike heat or pressure or a spark or even a hard rub of a finger will rouse to an explosion more violent than that to be produced from an equal quantity of dynamite or nitroglycerine. And all this, that hunter or soldier, marksman or police officer, may know that the crook of a finger will give him an instant mastery of time and space and accuracy and energy that only electricity itself can surpass.

And a Scots Presbyterian clergyman of the long ago first thought of it all! Rev. Alexander John Forsyth (1768-1843), 52 years minister of a remote sea-coast parish of burn and brae in old Aberdeenshire. In recognition of the genius of his discovery a magnificent bronze memorial tablet—placed in 1930, 87 years after his death—the joint gift of whole British regiments, arms-guilds, and a host of Scots and English sportsmen, stands today on the walls of the Tower of London, the only memorial in honor of any individual ever erected within the precincts of that eight-and-a-half centuries old structure that has well been called "the heart of the British empire!" Could earthly tribute farther go! Other achievements, fames and names have crowded the walls and aisles of Westminster Abbey with statues, busts and inscriptions; but in "the Tower" his remembrance stands without a rival! His Alma Mater, the University of Aberdeen, installed in 1931 a replica of the Tower tablet, in King's College, where he graduated in 1786.

* * *

To be strictly accurate, however, Forsyth did not at all "invent" the percussion-cap. He only made it possible by discovering that certain explosive compounds already somewhat known could be detonated by a blow and so used to ignite the charge of a gun . . .

The gray powder of his fulminate, which he called "detonating powder " Forsyth applied in half a dozen ingenious mechanisms to the firing of both "long" and "short" arms, military and sporting, and to cannon as well. It was, however, so scantily and unreliably made by chemists from whom he secured it that when summoned to the Tower of London to carry on his experiments for the Government he was compelled to make his own; for which he was ably qualified, having "majored" in chemistry at the University of Aberdeen. The delicacy and danger of the process, the handling of the product and the startling and powerful fumes accompanying its preparation, made the Government workmen flee the Tower as from a volcano, when he was at it in his laboratory!"

(For clarity of detail, all primers are illustrated twice their actual size)

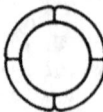

Two types of percussion caps for both long and short arms. The one at the left is known as the "hat type" due to its resemblance to the old stove pipe hats.

An unusual percussion cap in which the fulminate was contained in concave depression on the outside of the cap. The cap itself fitted on the hammer of the gun instead of on the nipple as was the usual method.

Tube primer in which the fulminate was contained within a copper tube.

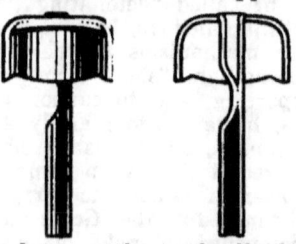

Referred to as the umbrella hooded tube primer due to appearance, this is one type of primer seldom encountered.

Called the mushroom primer. Another type of the hooded tube primers.

Two strips of paper between which were sealed small fulminate pellets formed a roll of Maynard tape primers.

Disk primers in which the fulminate was contained between thin copper foil disks, were used on some of the early breech-loading percussion arms. These were commonly known as Lawrence primers.

Allen's Patent primer. The fulminate was contained in the rim of the primer—quite like that of an ordinary rim fire cartridge.

Pin and tiny percussion cap for reloading pin fire cartridges.

Berdan primer—(Hobb's Pat.). Fulminate sealed in a brass cap by a thin sheet of thin silver foil paper.

Cross-section view of some early self-contained primers. These types, it will be observed, are different from the Berdan type of primer in that they contain the anvil as well as the fulminate.

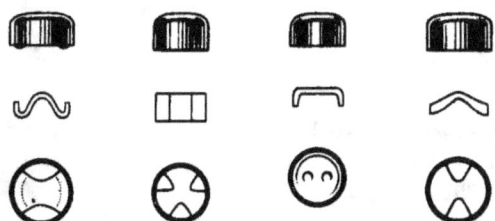

Illustrating four late types of self-contained primers. Top—side view of primers. Middle—cross-section side view of anvils. Lower—under side of primers with anvils in position.

BULLETS

There seems to be no end to the shape, sizes and styles in which bullets have been produced, or with which experiments have been conducted. In addition to the many types illustrated throughout the digest, the following are included . . . and reproduced actual size . . . for their interest. (weighed on druggist's scales).

.28 cal. ball for Paterson Colt revolvers—weight 36 grs.

.34 cal. ball for Paterson Colt revolver—weight 54 grs.

.31 cal. triangular for pistol—weight 60 grs.

.34 cal. hexagonal base lead bullet—weight 144 grs.

.31 cal. lead bullet for 10-shot Walch percussion revolver—weight 78 grs. The Walch is the arm in which two charges were loaded in each of the five chambers of the cylinder.

.36 cal. for Colt revolving rifle—weight 163 grs.

.36 cal. lead bullet for two-shot Lindsay percussion pistol—weight 125 grs. Two charges were loaded in the single barrel of the Lindsay — the rear charge serving as a cushion for the front charge.

.36 cal. composite 2-piece swaged bullet. Front half, hard lead . . . rear half, soft lead—weight 305 grs.

.37 cal. square lead bullet with slight twist for square bore rifle—weight 264 grs.

.38 cal. unknown—weight 286 grs.

.40 cal. lead bullet with copper tube . . . used in the 40-82 Winchester Express cartridge.

.41 cal. "Sugar Loaf" bullet for muzzle loading rifles . . . weight 187 grs. Also known as "Picket Bullets" . . . at least three sizes are known.

.41 cal. Unknown . . . weight 448 grs.

.42 cal. Base Band Pope type bullet for a false muzzle rifle . . . weight 240 grs.

.44 cal. bullet and ball for Whitneyville Walker revolver. Ball—weight 150 grs. Bullet—weight 230 grs.

.45 cal. lead bullet with copper tube . . . for the 45-90 W.C.F. Express cartridge.

.45 cal. cast in an Ideal Mould for the 45-70 cartridge . . . weight 520 grs.

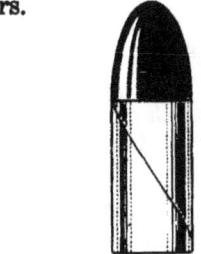 

.50 cal. lead paper patch bullet for the big fifty Sharps . . . weight 620 grs.

.50 cal. illustrated without the paper patch is this big 657 grain lead bullet for the big fifty Sharps cartridge.

.50 cal. lead bullet cast in a Mass. Arms Co. Mould . . . weight 505 grs.

.56 cal. lead bullet used in the 56 cal. Colt revolving rifle . . . weight 490 grs.

.56 cal. Jacobs four-grooved bullet used in the Jacobs four-grooved rifle . . . weight 747 grs.

.57 cal. English muzzle-loading rifle bullet . . . weight 530 grs.

.58 cal. three piece composite bullet used in some Civil War muskets. The saucer-shape disk is of tinned metal . . . the bullet and base being of lead.

.58 cal. famous Civil War "Minié ball" as cast on an Ideal Mould . . . weight 494 grs.

.60 cal. pointed lead bullet of unknown origin . . . weight 490 grs.

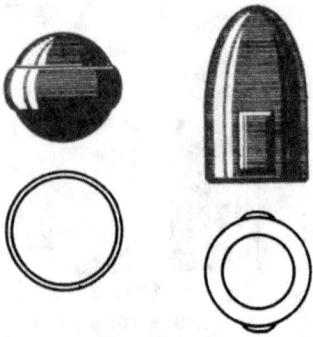

Belted ball and bullet for use in the two-grooved Purdy rifle of 1868 . . . ball weight, 470 grs. . . . bullet weight, 712 grs.

Eley's Destructor or Lethal ball, contains 16 small lead pellets . . . weight 490 grs.

Rifled lead slug for use in 12 ga. shotgun shell . . . weight 408 grs.

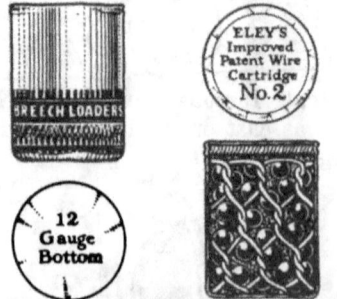

The forerunner of this shot container was patented (No. 5,570) in England in 1827. It was used in muzzle-loading shotguns.

.256 cal. Newton wire-protected tip.

.30-06 cal. Umbrella point.

.30-06 cal. Thomas patent point.

.35 cal. Hoxie bullet showing the round ball imbedded in the lead core of the metal jacketed, soft point bullet.

LIST OF CARTRIDGES
(For Small Arms)
PATENTED IN THE UNITED STATES
Prior to January 1, 1878
(From a book of patent papers issued by the Patent Office Branch Print in 1878)

Patent No.	Patentee	Date
5,699	Walter Hunt	Aug. 10, 1848
	Reissued Feb. 26, 1850, No. 163	
5,701	Walter Hunt	Aug. 10, 1848
	Reissued Feb. 26, 1850, No. 164	
7,147	A. D. Perry	Mar. 5, 1850
8,956	Marston and Goodell	May 18, 1852
11,496	Smith and Wesson	Aug. 8, 1854
11,870	Daniel Moore	Oct. 31, 1854
12,545	A. R. Davis	Mar. 20, 1855
12,556	A. N. Newton	Mar. 20, 1855
12,942	C. F. Brown	May 29, 1855
14,147	Smith and Wesson	Jan. 22, 1856
14,491	A. E. Burnside	Mar. 25, 1856
	Reissued Mar. 10, 1863, No. 1428	
15,141	Edward Maynard	June 17, 1856
	Reissued May 28, 1861, No. 1191	
15,369	Buckle and Dorsch	July 22, 1856
15,707	Julius Riedel	Sept. 9, 1856
15,996	Geo. W. Morse	Oct. 28, 1856
17,287	Edward Lindner	May 12, 1857
17,702	Gilbert Smith	June 30, 1857
	Reissued Sept. 14, 1858, No. 598	
17,792	W. B. Johns	July 14, 1857
18,143	J. D. Greene	Sept. 8, 1857
18,217	Lemuel Wells	Sept. 15, 1857
20,214	Geo. W. Morse	May 11, 1858
20,727	Geo. W. Morse	June 29, 1858
21,253	Gomez and Mills	Aug. 24, 1858
22,565	Edward Maynard	Jan 11, 1859
	Reissued May 28, 1861, No. 1192	
24,548	J. H. Ferguson	June 28, 1859
24,726	Ellis and White	July 12, 1859
	Reissued Aug. 25, 1863, No. 1529	
27,428	J. W. Cochran	Mar. 13, 1860
27,791	Geo. P. Foster	April 10, 1860
27,933	Smith and Wesson	April 17, 1860
	Reissued June 4, 1867, No. 2636	
29,080	B. B. Hotchkiss	July 10, 1860
29,108	Christian Sharps	July 10, 1860
	Reissued April 21, 1863, No. 1455	
29,287	J. P. Lindsay	July 24, 1860
30,109	Ethan Allen	Sept. 25, 1860
31,815	C. A. McEvoy	Mar. 26, 1861
32,345	Roberts Bartholow	May 21, 1861
32,949	Edward Lindner	July 30, 1861
	Reissued Feb. 17, 1863, No. 1411	
33,393	Johnston and Dow	Oct. 1, 1861
33,429	R. C. English	Oct. 8, 1861
33,481	J. P. Gillespie	Oct. 15, 1861
33,611	William Mont Storm	Oct. 29, 1861
33,805	Rollin White	Nov. 26, 1861
	Reissued Feb. 8, 1870, No. 3833	
34,061	Johnston and Dow	Jan. 7, 1862
34,367	Julius Hotchkiss	Feb. 11, 1862
34,579	Benedikt King	Mar. 4, 1862
34,615	Alexander Shannon	Mar. 4, 1862
34,713	E. C. Dunning	Mar. 18, 1862
34,725	Doremus and Budd	Mar. 18, 1862
34,744	Doremus and Budd	Mar. 25, 1862
34,806	B. L. Budd	Mar. 25, 1862
34,854	S. W. Wood	April 1, 1862
	Reissued April 2, 1872, No. 4843	
34,987	Christian Sharps	April 15, 1862
35,687	Johnston and Dow	June 24, 1862
35,699	J. C. Mayberry	June 24, 1862
35,872	W. H. Elliot	July 15, 1862
	Reissued Sept. 23, 1873, No. 5577	
35,878	Henry Kellog	July 15, 1862
	Reissued March 27, 1877, No. 7569	
35,949	E. O. Potter	July 22, 1862
36,066	Roberts Bartholow	Aug. 5, 1862
36,108	W. R. Pomeroy	Aug. 5, 1862
37,481	C. R. Alsop	Jan. 27, 1863
37,491	L. B. Bruen	Jan. 27, 1863
38,322	W. E. Moore	April 28, 1863
38,414	E. K. Root	May 5, 1863
39,109	William Bakewell	July 7, 1863
39,823	Edward Maynard	Sept. 8, 1863
39,869	J. H. Vickers	Sept. 8, 1863
39,915	Albert Hall	Sept. 15, 1863
40,092	W. H. Dibble	Sept. 29, 1863
40,111	Edward Maynard	Sept. 29, 1863
40,112	Edward Maynard	Sept. 29, 1863
40,490	W. W. Marston	Nov. 3, 1863
40,761	O. D. Lull	Dec. 1, 1863
40,978	Silas Crispin	Dec. 15, 1863
40,988	Rodman and Crispin	Dec. 15, 1863
41,183	David Williamson	Jan. 5, 1864
41,590	E. G. Allen	Feb. 16, 1864
41,684	George Conover	Feb. 23, 1864
42,329	Silas Crispin	April 12, 1864
42,388	Edward Maynard	April 19, 1864
42,666	Johnston and Dow	May 10, 1864
42,667	Johnston and Dow	May 10, 1864
42,668	Johnston and Dow	May 10, 1864
42,815	C. J. Bergen	May 17, 1864
43,851	J. C. Howe	Aug. 16, 1864
44,660	E. K. Root	Oct. 11, 1864
44,692	C. E. Sneider	Oct. 11, 1864
45,079	E. K. Root	Nov. 15, 1864
45,210	C. E. Sneider	Nov. 22, 1864
45,227	J. M. Connell	Nov. 29, 1864
45,292	W. H. Wills	Nov. 29, 1864
45,319	J. M. Cooper	Dec. 6, 1864
45,420	Edward Maynard	Dec. 13, 1864
45,666	Theodore Yates	Dec. 27, 1864
45,830	Samuel Jackson	Jan. 10, 1865

Patent No.	Patentee	Date
46,034	H. C. Spaulding	Jan. 24, 1865
46,292	Hiram Berdan	Feb. 7, 1865
47,317	D. F. Mefford	April 18, 1865
47,688	Ethan Allen	May 16, 1865
	Reissued May 1, 1877, No. 7647	
48,536	W. C. Dodge	July 4, 1865
	Reissued July 13, 1869, No. 3548	
48,820	Edwin Martin	July 18, 1865
49,237	Silas Crispin	Aug. 8, 1865
49,773	Orazio Lugo	Sept. 5, 1865
50,536	T. J. Powers	Oct. 17, 1865
	Reissued Sept. 14, 1869, No. 3638	
50,592	Jackson and Pusey	Oct. 24, 1865
51,243	William Tibbals	Nov. 28, 1865
51,324	T. T. S. Laidley	Dec. 5, 1865
51,672	T. P. Shaffner	Dec. 29, 1865
52,370	J. W. Smith	Jan. 30, 1866
52,818	Hiram Berdan	Feb. 27, 1866
53,168	Arthur Moffatt	Mar. 13, 1866
	Reissued July 19, 1870, Nos. 4075 and 4076	
53,388	Hiram Berdan	Mar. 20, 1866
53,490	W. H. Risley	Mar. 27, 1866
53,501	T. L. Sturtevant	Mar. 27, 1866
54,038	T. L. Sturtevant	April 17, 1866
55,233	A. S. Blake	June 5, 1866
55,552	T. L. Sturtevant	June 12, 1866
55,676	T. S. Laidley	June 19, 1866
58,800	G. A. Fitch	Oct. 16, 1866
59,044	Edward Maynard	Oct. 23, 1866
60,814	O. F. Winchester	Jan. 1, 1867
61,225	Edward Maynard	Jan. 15, 1867
62,283	I. M. Milbank	Feb. 19, 1867
	Reissued Aug. 6, 1867, No. 2716	
62,466	A. J. Bergen	Feb. 26, 1867
62,467	A. J. Bergen	Feb. 26, 1867
65,774	Dexter Smith	June 11, 1867
68,609	J. M. Crockett	Sept. 10, 1867
68,960	J. F. Cranston	Sept. 17, 1867
69,707	Jacob Rupertus	Oct. 8, 1867
70,612	Joseph Rider	Nov. 5, 1867
72,982	Thomas Cullen	Jan. 7, 1868
73,739	Henry Meigs, Jr.	Jan. 28, 1868
73,877	J. F. Cranston	Jan. 28, 1868
75,019	W. O. Howard	Mar. 3, 1868
78,337	William Tibbals	May 26, 1868
78,953	R. J. Gatling	June 16, 1868
81,058	Bethel Burton	Aug. 11, 1868
81,478	J. F. Cranston	Aug. 25, 1868
82,587	Hiram Berdan	Sept. 29, 1868
	Reissued Aug. 1, 1871, No. 4491	
83,434	Abraham and Bayliss	Oct. 7, 1868
85,482	Wilhem Schmitz	Dec. 29, 1868
87,125	William Tibbals	Feb. 23, 1869
87,297	B. S. Roberts	Feb. 23, 1869
87,352	J. V. Meigs	Mar. 2, 1869
87,735	J. R. Van Vechten	Mar. 9, 1869
87,990	G. H. Todd	Mar. 16, 1869
88,191	Edwin Martin	Mar. 23, 1869
88,202	W. F. Parker	Mar. 23, 1869
88,948	A. B. Ely	April 13, 1869
89,088	Smith and Storrs	April 20, 1869
89,563	G. H. Daw	May 4, 1869
90,607	William Tibbals	May 25, 1869
90,951	J. V. Meigs	June 8, 1869
91,278	Dexter Smith	June 15, 1869
91,668	Wesley Richards	June 22, 1869
91,818	E. M. Boxer	June 29, 1869
92,136	David Williamson	June 29, 1869
92,795	J. J. Chaudun	July 20, 1869
93,545	I. M. Milbank	Aug. 10, 1869
93,546	I. M. Milbank	Aug. 10, 1869
94,210	B. B. Hotchkiss	Aug. 31, 1869
96,373	Freidrich Wohlgemuth	Nov. 2, 1869
97,537	Logan and Eldridge	Dec. 7, 1869
97,615	Depew and Slatcher	Dec. 7, 1869
97,653	C. W. Lancaster	Dec. 7, 1869
97,843	Rollin White	Dec. 14, 1869
98,278	Leet and Hotchkiss	Dec. 28, 1869
98,439	W. H. Smith	Dec. 28, 1869
98,995	Sewell Newhouse	Jan. 18, 1870
99,078	Edwin Gomez	Jan. 25, 1870
99,079	Edwin Gomez	Jan. 25, 1870
99,528	F. E. Boyd	Feb. 8, 1870
99,666	Edwin Gomez	Feb. 8, 1870
99,721	W. H. Smith	Feb. 8, 1870
102,051	Oswald Schevenell	April 19, 1870
102,109	A. N. C. Gavard	April 19, 1870
102,675	R. J. Gatling	May 3, 1870
102,984	C. E. Sneider	May 10, 1870
103,079	T. J. Powers	May 17, 1870
104,912	N. G. Whitmore	June 28, 1870
105,348	Edwin Martin	July 12, 1870
108,543	D. E. Williams	Oct. 18, 1870
	Reissued July 3, 1877, No. 7783	
109,931	W. I. Page	Dec. 6, 1870
110,264	R. R. Moffatt	Dec. 20, 1870
110,265	R. R. Moffatt	Dec. 20, 1870
110,383	R. R. Moffatt	Dec. 20, 1870
110,881	Rollin White	Jan. 10, 1871
111,377	S. W. Paine	Jan. 31, 1871
111,856	Edwin Martin	Feb. 14, 1871
112,305	Rollin White	Feb. 28, 1871
113,634	Silas Crispin	April 11, 1871
113,677	C. W. Lovett, Jr.	April 11, 1871
115,498	I. M. Milbank	May 30, 1871
115,548	C. S. Wells	May 30, 1871
115,892	T. J. Powers	June 13, 1871
116,094	T. J. Powers	June 20, 1871
116,105	W. S. Smoot	June 20, 1871
116,640	C. E. Sneider	July 4, 1871
117,173	A. C. Hobbs	July 18, 1871
117,388	J. S. Crary	July 25, 1871
119,357	A. C. Hobbs	Sept. 26, 1871
120,323	G. R. Pierce	Oct. 24, 1871
120,338	W. S. Smoot	Oct. 24, 1871
120,403	G. R. Stetson	Oct. 31, 1871
120,529	Alwin Payne	Oct. 31, 1871
120,625	J. W. Cochran	Nov. 7, 1871
120,630	C. F. & J. E. DeDartein	Nov. 7, 1871
120,990	Henry Metcalfe	Nov. 14, 1871
121,606	Forehand & Wadsworth	Dec. 5, 1871
121,808	Alwin Payne	Dec. 12, 1871
122,399	I. M. Milbank	Jan. 2, 1872
122,504	C. S. Wells	Jan. 2, 1872
123,351	I. M. Milbank	Feb. 6, 1872
123,352	I. M. Milbank	Feb. 6, 1872
123,622	G. H. Dupee	Feb. 13, 1872
125,830	I. M. Milbank	April 16, 1872
126,058	W. W. Hubbell	April 23, 1872
127,308	J. W. Cochran	May 28, 1872
130,679	N. G. Whitmore	Aug. 20, 1872
131,016	I. M. Milbank	Sept. 3, 1872
131,017	I. M. Milbank	Sept. 3, 1872
131,018	I. M. Milbank	Sept. 3, 1872
131,104	A. D. Laws	Sept. 3, 1872
131,189	C. E. Sneider	Sept. 10, 1872
132,227	S. W. Wood	Oct. 15, 1872
134,048	Paul Giffard	Dec. 17, 1872
134,368	DeWitt C. Farrington	Dec. 31, 1872
136,130	Bethel Burton	Feb. 25, 1873
136,168	I. M. Milbank	Feb. 25, 1873
136,336	S. W. Paine	Feb. 25, 1873
	Reissued Nov. 10, 1874, No. 6130	
136,468	W. H. Tooth	Mar. 4, 1873
138,679	Mott and Gardiner	May 6, 1873
140,144	T. T. S. Laidley	June 24, 1873

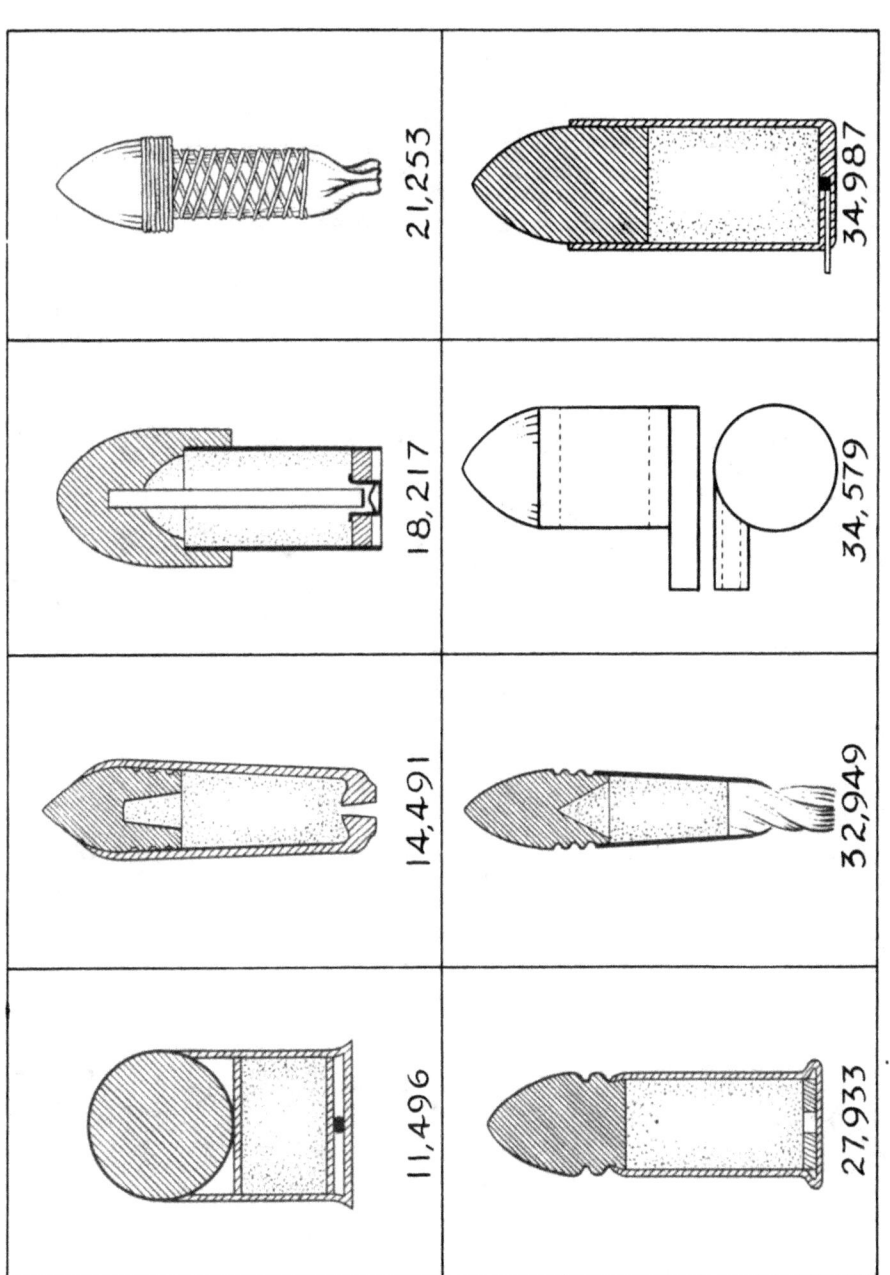

Cartridges (For Small Arms) Patented in the United States Prior to January 1, 1878.

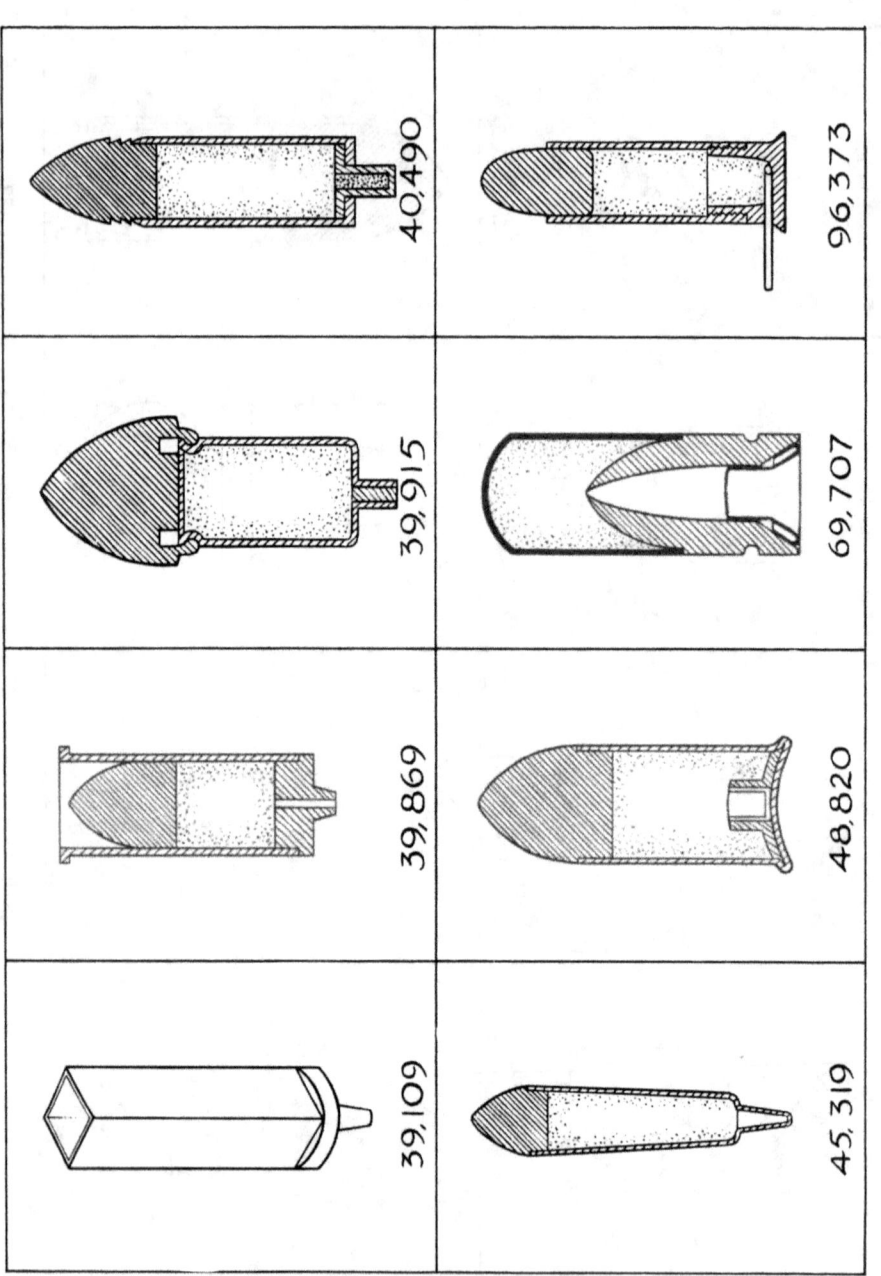

Cartridges (For Small Arms) Patented in the United States Prior to January 1, 1878.

Patent No.	Patentee	Date
142,924	Logan and Hart	Sept. 16, 1873
144,010	S. W. Wood	Oct. 28, 1873
144,011	S. W. Wood	Oct. 28, 1873
144,012	S. W. Wood	Oct. 28, 1873
147,871	Thomas Shaw	Feb. 24, 1874
148,467	A. B. Kay	Mar. 10, 1874
151,121	Hart and Logan	May 19, 1874
151,327	Charles Weldon	May 26, 1874
151,396	Edward Jones	May 26, 1874
152,428	W. S. Smoot	June 23, 1874
155,841	Jerome Orcutt	Oct. 13, 1874
157,793	J. W. Cochran	Dec. 15, 1874
157,916	B. B. Hotchkiss	Dec. 22, 1874
158,494	B. B. Hotchkiss	Jan. 5, 1875
159,665	M. M. E. Gauthey	Feb. 9, 1875
159,883	T. R. Bayliss	Feb. 16, 1875
160,263	DeWitt C. Farrington	Mar. 2, 1875
160,763	F. W. Freund	Mar. 16, 1875
161,514	N. C. Hunting	Mar. 30, 1875
162,901	E. J. Collett	May 4, 1875
163,154	R. F. Cook	May 11, 1875
163,181	J. H. Gill	May 11, 1875
164,894	B. D. Wilson	June 22, 1875
169,806	G. E. Hart	Nov. 9, 1875
169,807	G. E. Hart	Nov. 9, 1875
170,643	George Smith	Nov. 30, 1875
172,382	L. W. Broadwell	Jan. 18, 1876
172,446	A. B. and R. A. Kay	Jan. 18, 1876
172,714	L. T. DeFroidville	Jan. 25, 1876
173,538	Albert Hall	Feb. 15, 1876
175,293	Joseph Merwin	Mar. 28, 1876
175,400	Thomas Wilkinson	Mar. 28, 1876
178,055	Albert Hall	May 30, 1876
178,683	W. S. Smoot	June 13, 1876
178,698	S. W. Wood	June 13, 1876
179,634	Rollin White	July 4, 1876
180,510	Welsh and Evans	Aug. 1, 1876
180,840	H. C. Bull	Aug. 8, 1876
181,356	I. M. Milbank	Aug. 22, 1876
181,977	J. P. Pieri	Sept. 5, 1876
185,548	Israel Kinney	Dec. 19, 1876
185,835	Pierce and Eggers	Jan. 2, 1877
186,220	J. P. White	Jan. 16, 1877
186,391	J. P. White	Jan. 16, 1877
186,460	H. Buffington	Jan. 23, 1877
189,069	A. B. Smith	April 3, 1877
189,417	H. H. Barnard	April 10, 1877
190,190	B. L. Budd	May 1, 1877
190,208	J. H. Gill	May 1, 1877
191,243	Israel Kinney	May 29, 1877
191,430	B. B. Hotchkiss	May 29, 1877
192,676	J. H. Bullard	July 3, 1877
193,612	T. T. S. Laidley	July 31, 1877
193,658	B. B. Hotchkiss	July 31, 1877
193,855	Isaac Davis	Aug. 7, 1877
197,823	J. H. Bullard	Dec. 4, 1877
199,717	Jones and Marston	Jan. 29, 1878

LIST OF CARTRIDGES
Shown In
AMERICAN PATENTS FOR FIREARMS

(From a Digest of Cartridges for Small Arms by Bartlett & Gallatin, 1878)

Patent No.	Patentee	Date
1,422	B. F. Smith	Dec. 5, 1839
1,461	Hall and Day	Dec. 31, 1839
6,139	D. Minesinger	Feb. 27, 1849
11,191	E. Lindner	June 27, 1854
11,835	G. F. and A. H. Palmie	Oct. 24, 1854
12,567	A. T. Watson	Mar. 20, 1855
13,474	J. Swyney	Aug. 21, 1855
16,477	H. Genhart	June 27, 1857
22,348	E. Claude	Dec. 31, 1858
24,414	William Mont Storm	June 14, 1859
24,730	Gallager and Gladding	July 12, 1859
30,714	J. Boynton	Nov. 27, 1860
32,887	W. Palmer	July 23, 1861
36,531	B. S. Roberts	Sept. 23, 1862
36,571	M. Moses	Sept. 30, 1862
40,151	J. H. Wiechman	Sept. 29, 1863
40,572	Morgenstern & Morwitz	Nov. 10, 1863
40,992	J. W. Cochran	Dec. 22, 1863
46,131	F. D. Newbury	Jan. 31, 1865
50,125	C. Howard	Sept. 26, 1865
57,269	J. H. Selwyn	Aug. 14, 1866
58,525	D. Williamson	Oct. 2, 1866
62,465	A. J. Bergen	Feb. 26, 1867
72,803	L. Conroy	Dec. 31, 1867
74,594	S. S. Rembert	Feb. 18, 1868
74,888	C. Callaghan	Feb. 25, 1868
81,283	J. Merlett	Aug. 18, 1868
86,091	L. A. Merriam	Jan. 19, 1869
86,971	J. B. Conklin	Feb. 16, 1869
97,780	F. A. Le Mat	Dec. 14, 1869
105,093	J. Kraffert	Nov. 1, 1870
112,763	W. C. Dodge	Mar. 14, 1871
147,567	I. M. Milbank	Feb. 17, 1864

A PARTIAL LIST OF MANUFACTURERS HEADSTAMPS
...PAST AND PRESENT

A—American Cartridge Company
A (raised)—American Metallic Cartridge Co.
A (impressed)—American Eagle Cartridge Co. (on rimfires)
AA Co.—American Ammunition Co.
A.C.—Bologna Arsenal, Italy
Acorn—Gustave Genschow (RWS) Germany
AEP—Anciens Establissements Pieper (Belgium)
AL—Federal Cartridge Co. (Airline)
APC—Austin Powder Co.
ART—French Arsenal (government)
A&W—Allen & Wheelock
AWG—French manufacture
B—Braun & Bloem (Germany)
BB—Braun & Bloem (Germany)
Bird— " "
Bomb—Deutsche Waffen und Munitions Fabriken (Germany)
C. (raised)—Kynock
C. (impressed)—Creedmoor Cartridge Co.
CAC—Colonial Ammunition Co., Ltd., Aukland, New Zealand
CCC—Creedmoor Cartridge Co.
C.C.Co.—Clinton Cartridge Co.
CDL—C. D. Leet, Springfield, Mass.
C.O.A.—(or monogram of same) Industrias Quimicas Argentinas "Duperial" (Argentine)
CRB—Belgium manufacture

Cross—Switzerland
Crown—Sweden
C.T.M.—Crittenden & Tibbals Mfg. Co., (on rimfires)
D. (in a circle)—Gustave Genschow, Durlach Works (Germany)
D.—Dominion Cartridge Co. (Canada)
D—duPont Contract, World War I
D.D.C.—Dominion Cartridge Co.
D.C. Co.—   "   "   "
DA—Dominion Arsenal, Quebec
DAC—Dominion Ammunition Co. (Canada)
DEN—Denver Ordnance Plant, World War II
Diamond—Western Cartridge Co. (on rimfires)
DIANA—Gustave Genschow (shotshells)
DM—Des Moines Ordnance Plant
DMK—Deutsche Metalwaffen, Karlsruhe (Germany)
DWA—Deutsche Werke Aktiengesellschafft (Germany)
DWM—Deutsche Waffen und Munitions Fabriken (Germany)
E—Eley Brothers (England)
EB—   "   "   "
ELEY—Eley Brothers, (England)
EC—Evansville (Chrysler) Ordnance Plant
EC S—Evansville (Sunbeam) Ordnance Plant
EH—Hirtenberger (Austria)
EP—Montgomery Ward (on rimfires)
ENK—Greek Government manufacture
EW—Eau Claire Ordnance Plant
F—Frankford Arsenal (US)
F.A.—   "   "   "
F. (impressed)—Federal Cartridge Co. (rimfires)
FM—Fabrica Nacional de Municiones, Mexico, D. F.
F de M—Fabrica Nacional de Municiones, Mexico, D. F.
FN—Fabrique Nationale (Belgium) Browning Works
FNCM—Fabrica Nacional e Municoes SA, Sao Paulo, Brazil
G—French manufacture
G—Gambles Stores (rimfires)
G—   "   "   "
G. EGESTORFF—Linden . . . Hanover (Germany)
GF—Giulio Fiocchi (Italy)
GD—French manufacture
GECO—Gustave Genschow (Germany)
GD—Capua Arsenal (Italy)
GFL—Giulio Fiocchi (Italy)
GG (intertwined monogram)—Gevelot & Gaupillat (France)
GH—Griffin & Howe
G&S—Gambles Stores (shot shells)
G & Star—Gevelot & Gaupillat (Paris)
G & Star—Societe Francais Des Munitions (France)
Geneva Cross—Swiss National Armory
GR—G. Roth (Austria)
H—Winchester Repeating Arms (rimfires)
H (raised)—Winchester Repeating Arms (rimfires) early manufacture
H and 2, 3, or 4 stars—Hirtenberger Patronen Fabrik (Austria)
H and Date—Hirtenberger Patronen Fabrik (Austria)
HB—Houllier & Blanchard (Paris)
H.E. 1913-(Hirtenberger Patronen & Metalwarenfabrik Hirtenberg (Austria) on 6.5 Greek Mannlicher

HL—Haerins Krudtvaerk (Denmark)
HP—Federal (rimfires)
HIPOWER—Federal (shot shells)
ICI—Imperial Chemical Industries (England)
ICIANZ—Eley (Imperial Chemical Industries) plants in Australia and New Zealand
JG—Jacob Goldmark (rimfires)
K—Kynoch Cartridge Co. (England)
KC—Kynoch Cartridge Co.
KYNOCK—Kynoch Cartridge Co.
KB**—Austrian Manufacture
K&C—Kellar & Co. (Austria)
KN—Kings Norton Metal Co. (England)
L.—Lowell Cartridge Co.
LEON BEAUX & CO., MILANO (Italy)
LC—Lake City Ordnance Plant
Leaf—Keystone Cartridge Co. (Keenfire) . . . (Germany)
L
M—Hungarian Manufacture
MALLARD—Sears Roebuck & Co. (shot shells)
MFA CO. POINTER—Meriden Firearms Mfg. Co., Meriden, Conn.
MAXIM—Maxim Munitions Co., World War I
MW—Montgomery Ward (rimfires)
MONARK—Federal Cartridge Co. (shot shells)
N. . . in a shield—Rheinish Westphalische Sprengsdorff Exp. Co. (Germany)
N—Nobel (England)
NA CO.—Newton Arms Co.
NC—National Brass & Copper Co. World War I manufacture
Oak Leaf—Gustave Genschow (Germany)
ORBEA—Industrias Quimicas Argentinas 'Duperial'
OYJ—French Manufacture
P. (raised—Phoenix Cartridge Co.
P. (impressed)—Peters Cartridge Co.
PC CO.—   "   "   "
PETERS—   "   "   "
PARKER BROS. & CO., West Meriden, Conn. (shot shells)
PFC—Argentine Government
R. . . (in shield)—RWS (Rheinish Westphalische Sprengstoff, Germany)
RA—Remington Union Metallic Cartridge Co.
RaUMC—   "   "   "   "   "
RemUMC—   "   "   "   "   "
RAH—Remington Plant at Hoboken
RAS—Remington, made in the old Robin Hood plant at Swanton, Vermont
RED HEAD RELIANCE—Montgomery Ward (shot shells)
RHA CO.—Robin Hood Ammunition (Powder) Co., Swanton, Vermont
RHP CO.—Robin Hood Ammunition (Powder) Co., Swanton, Vermont
ROBIN HOOD—R o b i n Hood Ammunition (Powder) Co., Swanton, Vermont
Rising Sun Flag (impressed)—Japanese (shot shells)
R.L.—Royal Laboratory (England)
RMS—Rhenische Metalwaffen & Maschinenfabriken (Germany)
RWS—Rhenische Westfalische Sprengstoff (Germany)
RR CO.—Ross Rifle Co. (Canada)
SA CO.—Savage Arms Corp.
SA CORP.—Savage Arms Corp.
SAVAGE—Savage Arms Corp.
SAW—Sage Ammunition Works, Middletown, Conn.
SB—Sellier & Bellot (usually on pinfires)

SBP—Sellier & Bellot Plant (Prague, Czechoslovakia)
S&G—Gambles Stores (shot shells)
SFM—Societe Francaise des Munitions (France)
SKD—Keystone Cartridge Co., Keenfire (Germany)
SL—St. Louis Ordnance Plant
SP—Scorzato (Lujan, Argentine)
S.C.CO.—Southern Cartridge Co., Savannah, Ga. (Star Brand shot shells)
So. C. Co.—Southern Cartridge Co., Houston, Texas (Dixie, Retriever shot shells)
SMI—Italian Manufacture
SAS, or a bird in outline—Spreafico, Argentine
SAT—Suomen Ampumatarvetihdas (Finland)
Stars (4)—Hirtenberger, (Austria)
Superspeed—Winchester R. A. Co.
Super X—Western Cartridge Co.
Swastika—Chinese manufacture (1924)
TKY—Teikoku & Yakkyoseiza, Tokyo, Japan (shot shells)
Turkish Marks—(usually on 7.65 mm. Turkish Mauser)
TW—Twin City Ordnance Plant
U—Union Metallic Cartridge Co. (before merger with Remington)
UMC—Union Metallic Cartridge Co. (before merger with Remington)
U. (in a shield)—H. Utendoerffer
U—Utah Ordnance Plant
U HiSPEED—Remington UMC (on rimfires since World War II)
U.S.—United States Cartridge Co.
USC CO.—United States Cartridge Co.
US (raised)—United States Cartridge Co., early
US (intertwined like $ sign)—United States Cartridge Co.
VZ—German Manufacture
W—Winchester Repeating Arms Go., (9 mm. rim fire)
WRA—Winchester Repeating Arms Co.
WRA Co.—Winchester Repeating Arms Co.
Super Speed—Winchester Repeating Arms Co.
Stayness—Winchester Repeating Arms Co.
W. EVANS, LONDON—on rifle cartridges
W—Western Cartridge Co.
W Co.—Western Cartridge Co.
WCC—Western Cartridge Co.
WESTERN—Western Cartridge Co.
SUPER-X—Western Cartridge Co.
WR & Co.—Westley Richards (England)
XL—Federal Cartridge Co. (rimfires)
XR—Sears Roebuck & Co. (rimfires)
XTRA-RANGE—Sears Roebuck & Co. (shot shells)
1901 NEW RIVAL—Winchester Repeating Arms Co.
1901 LEADER—Winchester Repeating Arms Co.
1901 REPEATER—Winchester Repeating Arms Co.
1901 PIGEON—Winchester Repeating Arms Co.

AMERICAN AND CANADIAN AMMUNITION COMPANIES NOW OUT OF BUSINESS

American Ammunition Co., Oak Park, Ill. (shot shells)
American Buckle & Cartridge Co., West Haven, Conn. (shot shells)
American Cartridge Co., Kansas City, Mo.
Austin Cartridge Co., Cleveland, Ohio (shot shells)
Brown Standard Firearms Co., N. Y. (made a .45-100 2.6" PP, perhaps others)
Burnside Rifle Co., Providence, R. I. (made Burnside Metallic Percussion cartridges)
Gentral Cartridge Co., Kansas City, Mo.
Robert Chadwick, Hartford, Conn.
Chamberlain Cartridge & Target Co., Cleveland, Ohio
Charles F. Cook, (location unknown) (shot shells)
Clinton Cartridge Co. (trade name used by Western Cartridge Company to manufacture cartridges for Sears Roebuck)
Colt Cartridge Works, Hartford, Conn.
Crittenden & Tibbals Mfg. Co., South Coventry, Conn.
Delaware Cartridge Co., Wilmington, Del. (shot shells)
Eley Bros. Ltd., Transcona, Manitoba, Canada (shot shells)
B. C. English, Springfield, Mass. (Poultney's Pat. Paper & Brass and Maynard brass percussion cartridges)
Ethan Allen & Co., Worcester, Mass. (metallic)
E. Remington & Sons, Herkimer, N. Y.
Elam O. Potter, NYC (manufactured Johnston & Dow's Pat. Cartridges)
Fitch, Van Vechten & Co., NYC (made a .56-50 rim fire and others)
Hoxie Ammunition Co., Chicago, Ill.
Hazard Powder Co., Hazardville, Conn.
Harry B. Fisher, 7125 Woodland Ave., Philadelphia, Pa.
Hall & Hubbard (rimfires)
Hartsman Bros. & Allen (rimfires)
H. W. Mason Co., South Coventry, Conn. (Sharps Linen and Cup Primer types)
Jacob Goldmark, NYC (metallics)
John Krider, Philadelphia, Pa. (shot shells)
Johnston & Dow, 170 Broadway, NYC
Liberty Cartridge Co., Mount Carmel, Conn.
C. D. Leet, Springfield, Mass. (rim fire, metallics, shot shells)
Leet & Hall, Springfield, Mass.
Maxim Munitions Co., Watertown, N. Y.
Massachusetts Arms Co., Chicopee Falls, Mass.
Merrill Patent Firearms Mfg. Co., Baltimore, Md. (.56 Merrill and others)
New York Metallic Ammunition Co., Foot of 52nd St., NYC (.32 rim fire, perhaps others)
National Cartridge Co., St. Louis, Mo.
National Brass & Copper Tube Co. (location unknown) made .30-06's in 1918
National Arms Co., Brooklyn, N. Y. (teat fire)
New Haven Arms Co., New Haven, Conn.
National Projectile Works, Grand Rapids, Mich.
Newton Arms Co., Buffalo, N. Y.
Parker Bros., West Meridan, Conn. (shot shells)
Phoenix Cartridge Co., South Coventry, Conn. (rim fire, metallics)
Poultney & Trimble, Baltimore, Md.
Meriden Firearms Mfg. Co., Meriden, Conn. (shot shells)
Robin Hood Ammunition Co., Swanton, Vermont
Ross Rifle Co., Quebec, Canada
Sage Ammunition Works, Middletown, Conn.

Sage & Co., Middletown, Conn.
D. C. Sage, Middletown, Conn.
H. H. Schleber & Co., Rochester, N. Y.
Selby Smelting & Lead Co., California
Smiths Graphited Cartridge Co., Washington, D. C. (.45ACP, .30-06, shot shells)
Southern Cartridge Co., Houston, Tex., ('Dixie, 'Retriever' shot shells, .22 rim fire)
Southern Cartridge Co., Savannah, Ga. (Star brand shot shells)
Springfield Armory, USA., Springfield, Mass.
Standard Cartridge Co., Pinale, California
Smith & Wesson, Springfield, Mass.
Smith, Hall & Buckland, (location unknown) metallics
L. C. Siner & Co., (location unknown) shot shells
Sharps & Hankins, Philadelphia, Pa. (?)
Sharps Rifle Works, Hartford, Conn.
Union Cap & Chemical Co., East Alton, Ill. (.32 blanks, .22 shorts and others)
U. S. Cartridge Co., Lowell, Mass.
U. S. Cartridge Co., NYC
Volcanic Repeating Arms Co., New Haven, Conn. (Volcanic cartridges)
Worcester Cartridge Co., Worcester, Mass.
Watervliet Arsenal

RIM FIRE PISTOL AND RIFLE CARTRIDGES
Military and Sporting
(From UMC and Winchester catalogs of the '90's)

Description	U. M. C.		WINCHESTER	
	Powder Weight grains	Bullet Weight grains	Powder Weight grains	Bullet Weight grains
22 Short	4	29	3	30
22 Short, hollow point			3	27
22 Long	5	29	5	30
22 Colt's M. R. short	4	29		
22 Colt's M. R. long	5	29		
22 Long Rifle	5	40	5	40
22 Extra Long	6	40	7	40
22 Winchester	7	45	7	45
25	5	43	5	38
25 Stevens	10	67	11	65
30 Short	6	58	6	55
30 Long	8	75	9	55
32 Extra Short	5½	60	6	55
32 Short	9	80	9	82
32 Long	13	90	13	90
32 Long Rifle				
32 Extra Long	18	90	20	90
38 Short	15	125	18	130
38 Long	18	150	21	148
38 Extra Long	31	150	38	148
41 Short	10	130	13	130
41 Long	15	163	16	130
41 Swiss	58	300		
42 No. 64 Forehand & Wadsworth Rifle	20	220		
44 Short	15	200	21	200
44 Long	28	218	28	220
44 Henry Flat	26	200	28	200
44 Henry, pointed	26	200	26	200
44 C. & W.	26	200	23	200
44 Howard, long	33	218		
44 Extra Long	46	218	30	220
45 Danish	50	380		
45 Peabody				
46 Pistol, short	20	227	26	230
46 Rifle, long	35	305	40	300
46 Extra Long	57	305		
56/46 Spencer Sporting	45	320	45	330
50 Remington Pistol			23	290

Description				
50 Peabody	....	....	45	320
56/50 Spencer Carbine	45	350	45	350
50/70 Government	70	450	....	....
52 Carbine	70	400	....	....
52/70 Sharps	....	....	70	....
56/52 Spencer Rifle	45	500	45	386
56/56 Spencer Carbine	42	362	45	350
58 Springfield	60	500	....	....
58 Miller	....	....	....	....
58 Carbine	60	500	....	....
58 Gatling	70	575	....	....
58 Joslyn	44	380	45	350

CENTRAL FIRE PISTOL AND RIFLE CARTRIDGES
Military and Sporting

(From UMC and Winchester catalogs of the '90's)

Description	U. M. C.			WINCHESTER		
	Pri. No.	Pow. Wt. grs.	Bul. Wt. grs.	Pri. No.	Pow. Wt. grs.	Bul. Wt. grs.
22 Winchester S. S.	1	14	45	1	13	45
22 Extra Long	1	7	45	1	8	45
6 mm. U. S. Navy (.236)	....	....	....	....	....	112
25-20 Stevens & Win. S. S.	1	20	86	1	17	86
25-21 Stevens	1½	21	86	....	....	....
25-20 Marlin	1½	16	86	1	17	86
25-20 Single Shot	....	....	....	1	19	86
25-25 Stevens	1½	25	86	1	25	86
25-35 Win. Smokeless	....	....	....	5	25	117
25-35 Win. Short Range	....	....	....	5	35	86
25-36 Marlin	1½	....	86	5	36	106
30 Win. Smokeless Model 1894	....	....	....	5	30	160
30 Win. Short range	....	....	....	5	6	100
30 U. S. Army, Short range	....	....	....	2½	10	150
30 U. S. Army, Smokeless	....	....	....	2½	38	220
32 Protector	1	....	40½	1	4	40
32 Ideal	1	25	150	1	25	150
32 Short, Colt's	1	9	80	1½	9	82
32 Long Rifle	1	13	81	....	....	....
32 Long Colt's, Inside Lub.	1	13	83	1½	13	90
32 Colt New Police	1	13	100	1½	13	98
32 Smith & Wesson	0	10	88	1½	9	85
32 Smith & Wesson, Gallery	0	4	46	1½	4½	55
32 Smith & Wesson, Long	....	....	....	1½	13	98
32 Smith & Wesson Rifle	1½	17	98	1	17	100
32 Extra Long Ballard	1½	18	105	1	20	105
32 Winchester	1	20	100	1	20	115
32 Merwin & Hulbert, or H&R	0	15	88	1	15	88
32 Colt's Lightning Rifle	1	20	100	1	20	100
32-13 M & B, Short Range	1½	13	98	....	....	....
32-20 Marlin Safety	1	20	100	1	20	100
32-30 Remington Grooved	1½	30	125	....	....	....
32-40 Ballard	1½	40	185	2½	40	165
32-40 Remington Grooved	1½	40	150	....	....	....
32-40 Remington Patched	1½	40	150	....	....	....
32-40 Bullard Rifle	1½	40	150	1	40	150

32-44 Smith & Wesson Gallery	1	6	50	1½	4½	55
32-44 Smith & Wesson Target	1	11	83	1½	10	85
38 Smith & Wesson	0	15	146	1½	14	145
38 Smith & Wesson, Gallery	0	4	70	1½	4½	70
38 Merwin, Hulbert & Co.	0	15	142	1½	14	145
38 Short	1	15	125	1½	18	130
38 Long for Colt's D. A.	1	18	150	1½	19	150
38 Colt's Lightning Rifle	1	40	180	1	38	180
38 Winchester	1	40	180	1	38	180
38 Extra Long	1	31	146	1	38	160
38-20 M. & B. Short Range	1½	20	155	......	......	......
38-40 Marlin Safety	1	40	180	1	40	180
38-40 Remington Hepburn No. 3	1½	40	245	......	......	......
38-40 Remington Hepburn No. 3	1½	40	255	......	......	......
38-44 Smith & Wesson Gallery	1	6	70	1½	5½	70
38-44 Smith & Wesson Target	1	20	146	1½	20	146
38-45 Bullard Rifle	1½	45	190	2½	45	190
38-50 Remington Hepburn	1½	50	255	......	......	......
38-50 Remington Hepburn	1½	50	245	......	......	......
38-55 Ballard, Straight	1½	55	255	......	48	255
38-55 Marlin, Ballard & Win.	1½	55	255	2½	48	255
38-55 Short Range	......	......	......	2½	20	155
38-56 Winchester & Marlin	2½	56	255	2½	56	255
38-56 Colt's Lightning Rifle	2½	56	255	......	......	......
38-70 Winchester Model 1886	......	......	......	2½	68	255
38-72 Winchester Model 1895	......	......	......	2½	72	275
38 Express (38-90)	2½	90	217	2½	90	217
40-50 Sharps & Remington	B1	50	265	2½	50	285
40-50 Patched, Straight	......	......	......	2½	45	265
40-60 Winchester Rifle	1½	60	210	1	62	210
40-60 Marlin Mag. Rifle	1½	60	260	2½	60	260
40-60 Colt's Lightning Rifle	1½	60	260	......	......	......
40-65 Winchester Rifle	2½	65	260	2½	65	260
40-65 Sharps & Remington	2½	65	330	......	......	......
40-70 Sharps & Remington	2½	70	330	2½	65	330
40-70 Sharps, only, necked	B1	70	330	......	......	......
40-70 Winchester, Model 1886	2½	70	330	2½	70	330
40-70 Regular & Rem.	B1	70	365	......	......	......
40-70 Ballard, straight	1½	70	330	2½	70	330
40-70 What Cheer	B1	70	380	2½	70	380
40-70 Patched	......	......	......	2½	70	370
40-70 Bullard	......	......	......	2½	72	232
40-72 Winchester, Model 1895	......	......	......	2½	72	330
40-75 Bullard	1½	75	258	......	......	......
40-75 Winchester Express	2½	75	260	2½	75	260
40-82 Winchester Model 1886	2½	82	260	2½	82	260
40-85 Ballard	1½	85	370	2½	85	370
40-90 Sharps 2⅝" necked	B1	90	370	2½	90	370
40-90 Sharps 3¼"	2½	90	370	2½	90	370
40-90 Regular & Rem. necked	B1	90	365	......	......	......
40-90 Ballard 2 15/16	1½	90	370	2½	90	370
40-90 Bullard 2 1/16	1½	90	300	2½	90	300
40-90 What Cheer 2 1/16	B1	90	500	2½	90	500
40-110 Winchester Express	2½	110	260	2½	110	260
41 Short, Colt's Single Action	0	15	163	2	20	130
41 Short, Colt's Double Action	1	15	163	1½	14	160
41 Long, Colt's Double Action	1	22	196	1½	21	200
42 Russian	B1	77	370	2½	77	370
42 Russian Carbine	B1	60	370	2½	60	370

43 Spanish (patched 400 gr.)	B1	77	375	2½	77	395
43 Spanish Carbine	B1	60	375	2½	60	400
43 Egyptian	B1	70	400	2½	70	400
44 Smith & Wesson No. 3 Am.	1	23	218	2	25	205
44 Smith & Wesson No. 3 Rus.	2	23	246	2	23	255
44 Smith & Wesson Gallery	2	7	110	2	6	105
44 Smith & Wesson Grooved	2	7	110	2	6	115
44 Merwin, Hulbert & Co.	1	28	220	1½	30	220
44 Colt's	2	23	210	2	23	210
44 Long Ballard	1	35	227	....	....	....
44 Extra Long Wesson	1	50	257	....	....	....
44 Extra Long Ballard	1	50	265	....	....	....
44 Colt's Magazine Rifle	1	40	217	1	40	217
44 Winchester (44-40)	1	40	217	1	40	200
44 Evans New Model	2½	43	276	1	42	280
44 Evans Old Model	1	33	220	1	28	215
44 Webley	1	19	200	2	20	230
44 Bull Dog	1	15	170	2	15	168
44 Long	....	....	....	2½	48	250
44-40 Marlin Safety	1	40	217	1	40	217
44-60 Sporting, necked	B1	60	395	2½	60	395
44-77 Sharps 2½″	B1	77	405	....	....	....
44-77 Regular & Remington	B1	77	470	2½	77	470
44-90 Sharps 2⅝″	B1	90	500	....	....	....
44-90 Regular & Remington	B1	90	470	2½	90	470
44-90 Remington, special	B1	90	550	2½	90	520
44-90 Remington, straight	2½	90	550	....	....	....
44-95 What Cheer	B1	95	550	....	....	....
44-105 Sharps only, necked	B1	105	520	2½	105	520
45 Webley	1	20	230	2	20	230
45 Smith & Wesson	1	30	250	2	30	250
45 Colt's U.S.A.	2	40	250	2	35	255
45 Colt's U.S.A. (28 gr.)	2	28	250	2	28	255
45 Colt's Round Ball	2	7	138	....	....	....
45 Peabody-Martini	B1	81	480	....	....	....
45 Peabody-Martini Carbine	B1	55	405	2½	55	400
45 Government N.M. Solid Head	2½	70	500	....	....	....
45 Government Carbine	2½	55	405	2½	55	405
45 Government Solid Head	2½	70	405	....	....	....
45 Government Armory Practice	2½	5	140	2½	5	140
45 Martini-Henry	....	....	....	2½	85	480
45 Danish	B1	50	380	2½	50	380
45-15 Govt. Armory Practice	2½	15	230	....	....	....
45-20 Govt. Armory Practice	2½	20	230	....	....	....
45-35 Govt. Armory Practice	2½	35	365	....	....	....
45-50 Peabody Sporting	B1	50	290	2½	50	290
45-55 Government Carbine	2½	55	405	....	....	....
45-60 Winchester	2½	60	300	2½	62	300
45-60 Colt's Lightning Rifle	2½	60	300	....	....	....
45-70 Sharps, straight, 2 1/10″	B1	70	420	2½	70	420
45-70 Sharps Express 2 1/10″	B1	70	357	....	....	....
45-70 Marlin Magazine	1½	70	405	1	70	405
45-70 Government Patched Bullet	2½	70	345	....	....	....
45-70 Government Grooved Bullet	2½	70	345	....	....	....
45-75 Sharps, straight 2 1/10″	B1	75	420	2½	75	420
45-75 Winchester	2½	75	350	2½	75	350
45-82 Winchester	2½	82	405	2½	82	405
45-85 Winchester	2½	85	350	2½	85	650
45-85 Winchester Express	2½	85	300	2½	85	300

Cartridge						
45-85 Colt's Lightning Rifle	2½	85	290			
45-85 Bullard Rifle	1½	85	290	2½	85	295
45-85 Marlin	1½	85	285	1	85	285
45 Van Choate	B1	70	420	2½	70	420
45 Roumanian	B1	70	380	2½	70	380
45-90 Sharps, straight 2⁶⁄₁₀″	B1	90	500			
45-90 Winchester	2½	90	300			
45-90 Winchester Metal Patch				2½	90	295
45-105 Sharps, straight 2⅞″	B1	105	550			
45-120 Sharps, straight 3¼″	2½	120	550			
45-125 Sharps & Win. st. 3¼″	2½	125	500			
45-125 Winchester Exp. 3¼″	B1	125	300	2½	125	300
50 Pistol	B1	20	300	2	25	300
50 Government Carbine	2½	50	400	2½	50	400
50 Government Solid Head	2½	70	450			
50 Armory Practice	2½	8	200			
50-70 Government, sporting	2½	70	425	2½	70	425
50-70 Musket				2½	70	450
50-90 Sharps, straight	2½	90	473	2½	90	473
50-95 Winchester Express	2½	95	300	2½	95	300
50-95 Winchester, solid ball	2½	95	312			
50-95 Colt's New Lightning	2½	95	300			
50-95 Colt's N. L. solid ball	2½	95	312			
50-100 Winchester, Model 1886				2½	100	450
50-110 Winchester Express	2½	110	300	2½	110	300
50-115 Bullard Express	2½	115	300			
50-115 Bullard Solid Ball	2½	115	346			
57 Snyder	B1	80	480			
58 Transformed Musket	B1	80	530	2½	85	530
58 Carbine	B1	40	530	2½	40	530
58 Roberts	B1	60	620	2½	70	480
58 Snider, Turkish Model				2½	85	480

SHOT CARTRIDGES

(As listed in Hartley & Graham's Catalog No. 46)

1897

Rim Fire
- 22 B.B. Cap
- 22/50 Long
- 32 Long
- 38 Short
- 38 Long
- 44 Long
- 44 Henry
- 44 Short
- 56/50 (Spencer)
- 56/52   "
- 56/56   "

Central Fire
- 22 Winchester
- 32 Smith & Wesson
- 32 Short Colt's
- 32 Long Colt's
- 32 Merwin, Hulbert & Co.
- 32 Winch. Marlin & Colt's, L.M.R.
- 32 Extra Long Ballard
- 32/40 Winchester & Marlin
- 38 Smith & Wesson
- 38 Short Colt's
- 38 Long Colt's
- 38 Extra Long
- 38 Winch., Marlin & Colt's, L.M.R.
- 38/55 Winchester & Marlin
- 38/56   "   "
- 40/65   "   "
- 40/82   "   "
- 41 Long Colt's
- 44 Webley
- 44 Evans N. M.
- 44 Long Ballard
- 44 Merwin, Hulbert & Co.
- 44 Smith & Wesson Russian
- 44   "   "   American
- 44 Winch., Marlin & Colt's, L.M.R.
- 44 Extra Long Wesson

44 XL
45 Colt's
45 Smith & Wesson
45/40 Marlin
45/60 Winchester
45/70 Government
45/75 Winchester

45/90 Winchester
50 Pistol
50/50 Government Carbine
50/70 Government
50/95 Winchester
50/110 Winchester
300 Rook

PRIMERS
(Data from Winchester and UMC catalogs of the late '90's)

Winchester Primers

No. 1 Winchester Improved Primers, adapted to .32 Ex. Long, .22, .32, .38, .40, and .44 Winchester, .38 Ex. Long, .44 Evans, and .45-70 Marlin.

No. 1W Winchester Improved Primers, for *Nitro Powders*.

No. 1½ Winchester Improved Primers, adapted to .32 S. & W., .32 Colt, .32 Long, .32 Short, .38 Short, .38 M. & H., .38 Long Colt D. A., .38 S. & W., 41 Colt D. A., .44 M. & H. Same size as No. 1, but thinner metal and more sensitive.

No. 1½ W Winchester Improved Copper Primers for *Nitro Powders*.

No. 2 Winchester Improved Copper Primers, adapted to .41, .44 Bull Dog, .44 Webley, .44 S. & W. Russ., .44 S. & W. Am., .44 Colt, .45 Webley, .45 Colt, .45 S. & W., .50 Pistol, and Brass Shot Shells.

No. 2½ Winchester Improved Primers, adapted to all military and sporting cartridges.

No. 2½W Winchester Improved Primers for *Nitro Powders*.

No. 3W Winchester Improved Copper Primers for *Nitro Powders*.
This primer is the same in system as our celebrated No. 2 Winchester Primer, but is loaded with a priming mixture suitable for igniting Nitro or Black powders, and is marked "W." It will be found quick and sure fire when used with either Nitro or Black Powders.

No. 4 Winchester Primer for Winchester 'Leader' and 'Metal Lined' Shells for *Nitro Powders*.

No. 5 Winchester Improved Primer, made especially for .25-35 Winchester and .30 Winchester Smokeless Cartridges.

No. 6 Winchester Primer for Winchester "Repeater" Paper Shells for *Nitro Powders*.

We desire to call special attention to our Primer. The anvil in this is made from sheet brass, and has all the advantages of being pointed like the Berdan, with the additional advantage that, being a part of the primer, it is replaced at each reloading. When the anvil is made a part of the cartridge shell and of brass, it is blunted by the successive blows of the firing pin, and is worn out while the rest of the shell is yet serviceable. To prevent this, iron and steel anvils are often used, which batter and blunt the point of the firing pin.

The primer now offered is free from all these defects. Its anvil is sufficiently rigid to insure its firing, is renewed at every shot, and does not injure the gun in which it is used.

Berdan Primer No. 1 for Sporting and Military Cartridges
Berdan Primer No. 1½ for Metallic Shot Shells
Berdan Primer No. 2

U. M. C. Primers

No. 0   Copper
No. 1   Copper
No. 1½ Brass
No. 2   Copper, for "New Club" and Brass Shot Shells
No. 2½ Brass for Sporting and Military cartridges
No. 3   for "Trap," "Smokeless," and other High Grade U. M. C. Paper Shells.
No. 5   Copper for "Nitro" and High

No. 6　　Base Brand Paper Shells
　　　　Copper, for Smokeless Powder Cartridges
No. 6½　Brass, for Smokeless Powder cartridges
No. 7　　Copper, for Smokeless Powder Cartridges and "Primrose" Club Shells.
No. 7½　Brass, for Smokeless Powder Cartridges

No. 8½ Brass for High Pressure Smokeless Powder Cartridges with metal-cased bullets.
U. M. C. Primer No. 1, 1½, 6, and 6½ are the same size
U. M. C. Primer No. 2, 2½, 7, 7½ and 8½ are the same size
The U. M. C. Primers will interchange with same numbers of other makers

COMPARATIVE DIMENSIONS
For Cartridge Measurements

Millimeters to decimals of an inch

2.7 mm.	.106	7.5 mm.	.295	10.4 mm.	.409
3. mm.	.118	7.62 mm.	.300	10.75 mm.	.423
4. mm.	.157	7.63 mm.	.300	11. mm.	.433
4.25 mm.	.167	7.65 mm.	.301	11.15 mm.	.439
5. mm.	.197	7.7 mm.	.303	11.2 mm.	.441
5.5 mm.	.214	7.92 mm.	.312	11.43 mm.	.450
5.7 mm.	.224	8. mm.	.315	12. mm.	.472
6. mm.	.236	9. mm.	.354	12.7 mm.	.500
6.35 mm.	.250	9.3 mm.	.366	13. mm.	.512
6.5 mm.	.256	9.5 mm.	.374	14. mm.	.551
7. mm.	.276	9.8 mm.	.386	15. mm.	.591

Fractions of an inch in decimal equivalents

1/64	.0156	23/64	.3593	45/64	.7031
1/32	.0312	3/8	.375	23/32	.7187
3/64	.0468	25/64	.3906	47/64	.7343
1/16	.0625	13/32	.4062	3/4	.750
5/64	.0781	27/64	.4218	49/64	.7656
3/32	.0937	7/16	.4375	25/32	.7812
7/64	.1093	29/64	.4531	51/64	.7968
1/8	.125	15/32	.4687	13/16	.8125
9/64	.1406	31/64	.4843	53/64	.8281
5/32	.1562	1/2	.500	27/32	.8437
11/64	.1718	33/64	.5156	55/64	.8593
3/16	.1875	17/32	.5312	7/8	.875
13/64	.2031	35/64	.5468	57/64	.8906
7/32	.2187	9/16	.5625	29/32	.9062
15/64	.2343	37/64	.5781	59/64	.9218
1/4	.250	19/32	.5937	15/16	.9375
17/64	.2656	39/64	.6093	61/64	.9531
9/32	.2812	5/8	.625	31/32	.9687
19/64	.2968	41/64	.6406	63/64	.9843
5/16	.3125	21/32	.6562	1	1.0000
21/64	.3281	43/64	.6718		
11/32	.3437	11/16	.6875		

AMERICAN COUNTERPARTS OF FOREIGN CARTRIDGES

American	Foreign
.22 WCF (Hornet)	5.6x35 mm. Vierling
.22 Savage HiPower	5.6x52R
.25 Automatic Pistol	6.35 mm.
.25-35 WCF	6.5x52R
6.5 Mannlicher	6.5 mm. Mann. Schoen
7 mm.	7x57 mm.
.30-30 WCF	7.62x51R
7.62 Russian	7.62x63 mm.
.30 Mauser Pistol	7.63 mm.
.30 Luger Pistol	7.65 mm. Parabellum
.32 Auto Pistol	7.65 mm.
8 mm. Mauser	7.9, 7.92, or 7.9x57 mm.
.380 ACP	9 mm. Browning, Short
.38 S&W	.38 or .380 Webley & Scott
9 mm. Luger	9 mm. Parabellum
.404 British	10.75x73 mm.
.450/400 British	10.3x60 mm.

D W M MARKINGS

The DWM (Deutsche Waffen und Munitions Fabriken ... Germany) used a number instead of marking the caliber on cartridges. The following are the ones most commonly encountered ...

Cartridge	No.
5 mm. Bergmann Auto Pistol	416A
5 mm. Clement Auto Pistol	484
5.6 x 35R mm. Vierling (.22 WCF)	539
5.6 x 52R mm. (.22 Hi Power)	545
6 x 58 mm.	489
6.35 mm. (.25 Auto Pistol)	508A
6.5 x 27P mm.	476
6.5 x 48R mm.	463A
6.5 x 52 mm.	519A
6.5 x 52R mm.	519
6.5 x 53 mm.	394
6.5 x 53R mm.	395C
6.5 x 54 mm. (short, rimless)	457A
6.5 x 55 mm. (Mauser rifle)	431C
6.5 x 57 mm.	404A
6.5 x 58 mm. (Mauser rifle)	457
6.5 x 58R mm. (Mauser, rifle rimmed)	463
6.5 x 61 mm.	431L
6.5 mm. (Dutch Mannlicher Rifle, M93)	395D
6.5 mm. Italian Carcano rifle	473
6.5 mm. Japanese Arisaka rifle	481
6.5 mm. Bergmann Auto Pistol	413A
6.5 mm. Mannlicher-Schoenauer rifle	477
7 mm. Mauser rifle, rimmed—old model	M93B
7 x 57 mm. (Mauser rifle, rimless)	380L
7 x 57R mm. (Mauser rifle, rimmed)	380J
7 x 64 mm.	557
7 x 65R mm.	557A
7 x 72R mm.	573
7.62 x 51R mm. (.30-30 WCF)	543
7.62 x 63 mm. (30-06)	379E
7.63 mm. (.30) Mauser pistol	403
7.65 mm. (.32) Auto pistol	479A
7.65 mm. (.30 Luger) Parabellum pistol	471
7.65 mm. Mannlicher Auto pistol	466
7.65 mm. Mannlicher carbine	497
7.65 mm. Argentine Mauser rifle	367
7.7 mm. (.303) Lee Metford (British) rifle	453
8 x 48R mm. S&S	462A
8 x 50R mm. (M95 Mannlicher rifle)	358C
8 x 51 mm. Mauser M88 rifle, short model	366L
8 x 51R mm.	366L2
8 x 57R mm.	366B
8 x 57JR (Mauser rifle, short, rimless)	366D1
8 x 58R mm. S&S	462
8 x 57 mm. (Model 98 Mauser rifle, 7.92)	560
8 x 60 mm.	542
8 x 60R mm.	542A
8 x 64 mm.	553
8 x 65R mm.	553A
8 x 72R mm.	574
8 x 75 mm.	514A
8 x 75R mm.	514
8 mm. Swiss Schmidt-Rubin rifle	388
8 mm. Danish Krag rifle	358A
8 mm. Bergmann Simplex Auto pistol	488
8.15 x 46R mm.	455
8.2 x 66 mm. Mannlicher Schoenauer rifle	523
9 mm. Luger (Parabellum) Auto pistol	480C
9 mm. Luger (Parabellum) carbine	480D
9 mm. Steyr Auto pistol	577
9 mm. Mauser Auto pistol	487
9 mm. Browning Short (.380) Auto pistol	540
9 mm. Bergmann Auto pistol	456B
9 x 56 mm. (Mannlicher rifle)	491E
9 x 57 mm. (Mauser rifle)	491A
9 x 57R mm. Mauser rifle, rimmed	491B
9 x 70R mm.	474B
9.3 x 62 mm. (Mauser rifle)	474C
9.3 x 72R mm.	77D
9.3 x 74R mm. (Mauser Magnum rifle)	474A
9.5 x 57 mm. (Mannlicher-Schoenauer rifle)	531
10.75 x 63 mm.	515
10.75 x 68 mm.	515A
10.75 x 73 mm. (.404 Jeffery)	555
10.75 x 70 mm. (Mauser Magnum rifle)	495
11.15 x 60R mm.	41
11.2 x 60 mm. (Mauser rifle, M71)	41A
11.75 x 88E mm.	533

ADDENDUM

ADDITIONAL DATES FOR THE CARTRIDGE CHRONOLOGY

DATE	EVENT
1818	Joseph Manton of England developed and patented the tube primer. Approximately a sixteenth of an inch in diameter, and five-eighths of an inch in length, the tube contained fulminate of mercury. When laid lengthwise in a horizontal vent and struck by the hammer it detonated the powder charge in the barrel. The tube primer was used in the Merrill gun of that time, 14,500 of which were bought by the British Government. Manton's patent of 1816 had to do with a form of pellet detonation.
1824	Dreyse made copper percussion caps in Sommerda.
1855	The British Government contracted with Christian Sharps for 6,500 of his .52 caliber carbines. These were for the Sharps linen cartridge and were equipped with Maynard tape-primer magazines.
1890	Krag cartridge was designed in this year by Frankford Arsenal. The cartridges were ready before the actual gun model had been selected.
1895	The U. S. Navy adopted the 6mm Lee Straight Pull Bolt Action rifle. It was the smallest military rifle caliber ever used by the United States . . . and in fact the smallest used by any nation at the time of its adoption. It was also the first U. S. design of a clip-loading rifle.
1898	F. W. Olin founded the Western Cartridge Company.
1903	Introduction of the U. S. Springfield '03 rifle and the .30 caliber rimless cartridge for use in it.
1914	First use of progressive-burning powder in center fire ammunition by Winchester.
1922	First use of progressive-burning powder in shot shells by Western Cartridge Company.
1932	Purchase of the Winchester Repeating Arms Company by the Western Cartridge Company.
1944	Both Winchester and Western became divisions of Olin Industries, Inc.

ADDITIONAL DATA FOR SPECIMENS ILLUSTRATED

page 43 . . . (8mm) . . . The label on a box of these self-contained cartridges reads as follows . . .

<blockquote>
Capsules a balles Systeme Loron
Pour Pistolets et Carabines
de Salon
100 #6   G. D.
Gevelot   1835
</blockquote>

page 65 . . . (31 Unknown) . . . from the "Abridgements of Specifications, Small-Arms, 1889-9" . . . published by the British Patent Office. "#18,370 . . . Morris, R. Nov. 14 . . . Adapting for miniature ammunition practice. The miniature cartridge is shown in Fig. 2, and the dummy or adapter which receives it,

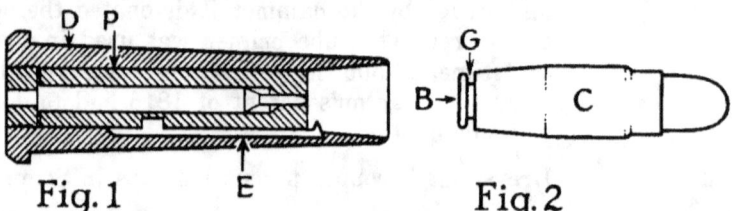

Fig. 1   Fig. 2

and which fits the chamber of the gun, is shown in Fig. 1. The invention is specially applicable for use with small-bore rifles. The plug P in the shell D carries the extractor claw E. The rear of the cartridge case C if grooved at G to form a flange B which is notched to permit the passage of the claw E, to enable the cartridge to be attached to the dummy by a bayonet joint." It is possible that the "31 Unknown" was used in such an adapter.

page 88 . . . (.307 Triangular Pistol) . . . Martin Retting turned up these triangular items around 1933. He found them with a Smith & Wesson revolver, which had been rebarreled and a new cylinder made. This experiment was the product of an ingenious German gunsmith of Brooklyn, N. Y. He had designed and built the tools for producing the cartridges. This gunsmith also produced a rifle barrel which used a square bullet.

page 124 . . . (9mm Luger) . . . The original German 9mm had the cone-type bullet. It was adopted by the German Navy this way in 1905 and by the Army in 1908. The cartridge is referred to as the "P-08". In 1915 the cone bullet was discontinued and the round nose adopted . . . however the cone bullet was made in this country until the early 1920's.

page 124 . . . (9mm Ultra) . . . According to Phil Sharpe, who discovered this experimental cartridge, there was but one lot of them made in 1940.

DATE	EVENT

page 124 . . . (357 Magnum) . . . Philip B. Sharpe, noted arms and ammunition technician, was personally responsible for this powerful cartridge. It was a private, non-profit transaction. Douglas B. Wesson of Smith & Wesson developed the gun while Phil developed the cartridge, using a variety of special and hand made cases. The gun and the cartridge were designed to operate at *rifle* pressure (factory standards are 38,000 pounds). DuPont, Hercules, Remington and Winchester all cooperated in the development.

page 175 . . . (6mm Perfumed) . . . This perfume cartridge was an RWS loading, once made for French trade. It is said to have been used in the house after a session of shooting with Flobert pistols . . . "to make the room smell good again."

page 185 . . . (58 cal. three piece composite bullet) . . . Col. B. R. Lewis writes as follows . . . "The 3-piece ball is Williams Pt. Variously issued, according to official circulars, in ratio of from 1 to 6 per packet of ten .58 caliber cartridges. It was to remove powder fouling . . . the zinc washer being expanded into the rifling when the pressure forced the base plug forward. These cartridges can be spotted easily as the nose of the bullet is considerably more rounded than the solid ball. Most all packets of .58's have one or more of them."

ADDITIONAL HEADSTAMPS

A. B. C. Co.—American Buckle and Cartridge Co.
A. B. C.—
A. C. Co.—Austin Cartridge Co., Cleveland, Ohio
A. J.—Alton Jones
ALOUETTE—
AMERICAN EAGLE—American Cartridge Co., Kansas City, Mo.
Anchor (raised) in circle——
A. P. C.—German make for Sears & Roebuck
B (Impressed)—Birmingham Metal & Munitions Co. (Birmingham Small Arms)
BALLISTIC TOOLS INC.—Middletown, Conn.
B. & B. D.—Braun & Bloom, Dusseldorf
BE MPLS—B. & E. Cartridge Co., Inc., Minneapolis, Minn.
BESCHUSS—(Test Loads) Germany
BRENNEKE-LEIPZIG—Germany
CBC—Sao Paulo, Brazil
C. F.—Carbine F. A. (Frankford Arsenal)
D & C—Dreyse & Collenbusch, Sommerda, Germany
DG—France
DI—Defense Industries Ltd., Verdun, Quebec
DOMINION—Dominion Cartridge Co., Montreal, Canada
D. R. Wz—
ECP P—Ecole Centrale de Pyrotechnie de Bourges, France
E. REMINGTON & SONS—
FA HP—Frankford Arsenal Hi Pressure Test
FA NM—FA (National Match)
FA RG—FA (Rifle Grenade)
FABRIO de GEVELOT PARIS—Gevelot et Cie, Paris, France
F. D. & Co.—F. Draper & Co.
F. GY; BP.F.P.—Fegyvergyer, Budapest
FIOCCHI—Giulio Fiocchi, Lecco, Italy
Flying Bird—Spreafico, Argentine
FOWLER—(brass shot shell)
FUSNOT BRUXELLES — Fusnot, Brussels, Belgium
F. V. V. & Co.—Fitch Van Vetchen & Co., New York City
GARDNER MACHINE GUN CO.—England
G. B.—Greenwood & Batley, Leeds, England
G. J.—(same as SFM)
GECADO—(Gustav Genshow)
GUSTAV GENSHOW—branch of RWS at Durlach
GYTTORP—Holland
HA—Haerens Ammunitions Arsenal, Denmark
HOLLAND & HOLLAND—
H. SEARS & CO.—
JOYCE & CO. LONDON—(shot shells)
KI—Kirkee Arsenal, India
KOLN-ROTTWELL — Rottweil Aktiengesselshafft, Cologne, Germany

M—(in 41 rf only) produced for Philip Jay Medicus by Rem-UMC
M—Milwaukee Ordnance Plant, Milwaukee.
Maltese cross with target in center—Union Cap & Chemical Co., name changed later to Western Cartridge Co.
MFA—Swiss
M. F.—Factory No. 1
M. F. ST. ETIENNE—Manufacture Armes de St. Etienne, France
MG—Factory No. 2, Footscray, Australia
$\frac{M}{W}$—Montgomery Ward
NORMA—A. B. Norma Amottors, Sweden
Pk — Zaklady Amunicyjne Pocisk, Spolka Akcyjna w Warsawie, Poland
PS—Povaske Strojarne Povaska Bystrica, Czechoslovakia
RF—FA (Rifle)
RG—Radway Green, England
RL—Lowell Cartridge Co. (Rifle)
RW—Winchester Repeating Arms Co. (Rifle)
SAKO—Finland
SAS (raised in circle)——Spreafico, Argentine
SCHERMER KARLSRUNE—Germany
SECOND QUALITY—Worcester Cartridge Co., Worcester, Mass.
J. B. SMITH—J. Bushnell Smith, Middlebury, Vermont
SP—Scorzato, Lujan, Argentine
SPEER—Vernon D. Speer, Lewiston, Idaho (300 Weatherby Magnum)
STANDARD ARMS CO.—Standard Arms Co., Wilmington, Del.
T. C. AFA—Askeri Fabriklar, Mamulati, Turkey
US 1776-1876 (Exhibition production 1876, Philadelphia Centennial, F. A.)
VC—Canadian Gov't Ordnance Plant, Verdun, Quebec
VE—French Gov't Cartridge Factory, Valence, France
V. F. M. & CA LIEGE—Capsulerie Leigoise Francotte May & Cie, Liege, Belgium
VON LENGERKE & DETMOLD—Von Lengerke & Detmold, New York City
VPT—Valtion Metallitehtaat, Helsinki, Finland
VZ—Vereinigte Zunder u Kabelwerke A. G. Meissen, Germany
W—on 9mm shotshell and ball cartridges by Winchester Repeating Arms Co. (1921 to date)
W. S. M. Co.—Worcester Stamped Metal Co., Worcester, Mass.
X (raised)—Delaware Cartridge Co., Wilmington, Delaware
Z—Povaske Strojarne Povaska Bystrica, Czechoslovakia

www.ingramcontent.com/pod-product-compliance
Lightning Source LLC
Chambersburg PA
CBHW072006110526
44592CB00012B/1217